21005

MODERN INTERMEDIATE ALGEBRA

ALSO BY MARGARET F. WILLERDING

Elementary Mathematics: Its Structure and Concepts, Wiley, 1966

Willerding and Hayward, *Mathematics: The Alphabet of Science,* Wiley, 1968

MODERN INTERMEDIATE ALGEBRA

Margaret F. Willerding
San Diego State College

Stephen Hoffman
Bates College

JOHN WILEY & SONS, INC.   NEW YORK · LONDON · SYDNEY · TORONTO

Copyright © 1969 by John Wiley & Sons, Inc.

All Rights reserved. No part of this book may be reproduced by any means, nor transmitted, nor translated into a machine language without the written permission of the publisher.

10 9 8 7 6 5 4 3 2 1

Library of Congress Catalog Card Number: 79-84972

SBN 471 94669 9

Printed in the United States of America

PREFACE

The material in this book is designed for a three-unit, one-semester, course in intermediate algebra for students with a relatively weak high school preparation. It is designed to present topics in algebra for those students who will pursue higher mathematics training as well as for those students for whom it is a terminal course. We have attempted to provide extensive development of manipulative techniques with a large number of exercises together with constant attention to the structure of the theory. Although we feel that students at this level do not need and, actually, are not ready for detailed proofs of the theorems they are using, we do believe they should know of the existence of theorems and realize that proofs exist. We have included some simple proofs; however, most of the theorems used are stated without proof. The presentation is modern, and emphasis is on the algebraic structure of the ordered field of real numbers.

Chapters 1 and 2 deal with the manner in which real numbers are written and how the manipulative aspects of these numbers are based on the field axioms. Chapter 3 treats the algebraic operations possible on polynomial expressions and Chapter 4 extends this treatment to rational expressions. Chapter 5 offers a full discussion of first-degree equations and inequalities in one variable ending with a collection of word problems presented as applications.

Chapter 6 introduces the concept of a function as a set of ordered pairs. The concept of linear function is presented here. Chapters 7 to 9 examine complex numbers, quadratic functions, and quadratic equations. The technique of completing the square is used both for the quadratic function (providing an application to maximum and minimum problems) and in the development of the quadratic formula. The complex numbers are presented as an algebraic extension of the real number system.

Chapter 10 presents systems of linear and quadratic equations together with second- and third-order determinants. Determinants are used to find solution sets of systems of linear equations. The conics are presented here.

Chapter 11 is a treatment of sequences and series with emphasis on arithmetic and geometric sequences and series. Here we present the binomial theorem.

Chapter 12 introduces the exponential function and the logarithmic function is presented as the inverse of the exponential function. Extensive computation using logarithms is included here.

Adequate exercises are provided for each section and a chapter review is included at the end of each chapter. Answers to odd-numbered problems are provided at the back of the book. Answers to even-numbered exercises are available to the instructor in a separate teacher's manual.

Margaret F. Willerding
Stephen Hoffman

CONTENTS

CHAPTER 1 Real Numbers and Their Properties   1
  1.1 Sets and Symbols   1
  1.2 The Equality Relation   4
  1.3 The Real Numbers and Their Properties   6
  1.4 Prime and Composite Numbers   13
  1.5 Addition and Subtraction of Real Numbers   14
  1.6 Multiplication and Division of Real Numbers   19

CHAPTER 2 Exponents and Radicals   25
  2.1 Laws of Positive Integer Exponents   25
  2.2 Roots and Radicals   30
  2.3 Positive Rational Number Exponents   32
  2.4 Zero and Negative Rational Exponents   35
  2.5 Changing the Form of Radicals   37
  2.6 Operations Containing Radical Expressions   41
  2.7 Scientific Notation   44

CHAPTER 3 Polynomials   46
  3.1 Definitions   46
  3.2 Classification of Polynomials   47
  3.3 Addition and Subtraction of Polynomials   49
  3.4 Multiplication of Polynomials   52
  3.5 Special Products   54
  3.6 Division of Polynomials   57
  3.7 Factoring   59
  3.8 Factoring Binomials and Trinomials   62
  3.9 Factoring the Sum and Difference of Two Cubes   66

CONTENTS

CHAPTER 4 Rational Expressions — 68
- 4.1 Definitions — 68
- 4.2 Simplification of Rational Expressions — 68
- 4.3 Least Common Multiples — 70
- 4.4 Addition and Subtraction of Rational Expressions — 71
- 4.5 Multiplication of Rational Expressions — 77
- 4.6 Division of Rational Expressions — 79
- 4.7 Complex Rational Expressions — 82

CHAPTER 5 First Degree Equations and Inequalities — 87
- 5.1 Algebraic Equations — 87
- 5.2 Equivalent Equations — 88
- 5.3 Solving Equations — 89
- 5.4 Solving Equations for Specified Symbols — 93
- 5.5 Identities — 95
- 5.6 Inequalities — 97
- 5.7 Applications of Equations and Inequalities — 101

CHAPTER 6 Relations and Functions — 109
- 6.1 Cartesian Products — 109
- 6.2 The Cartesian Coordinate System — 115
- 6.3 Open Sentences in Two Variables — 117
- 6.4 Relations and Functions — 122
- 6.5 Functional Notation — 126
- 6.6 Linear Functions and Their Graphs — 128
- 6.7 Direct and Inverse Variation — 132

CHAPTER 7 Complex Numbers — 138
- 7.1 Extension of the Real Number System — 138
- 7.2 Addition of Complex Numbers — 139
- 7.3 Subtraction of Complex Numbers — 140
- 7.4 Multiplication of Complex Numbers — 142
- 7.5 Division of Complex Numbers — 144
- 7.6 Standard Form of a Complex Number — 147
- 7.7 The Conjugate of a Complex Number — 149
- 7.8 Square Roots Which are Complex Numbers — 151

CHAPTER 8 Quadratic Functions — 154
- 8.1 Definitions — 154
- 8.2 The Graph of a Quadratic Function — 155

8.3	Completing the Square	157
8.4	Applications	162

CHAPTER 9 Quadratic Equations and Inequalities — 168

9.1	Quadratic Equations	168
9.2	Solution Set of $x^2 = k$	168
9.3	Solution by Factoring	170
9.4	The Quadratic Formula	173
9.5	Properties of Roots of Quadratic Equations	176
9.6	Fractional Equations	179
9.7	Equations Containing Radicals	181
9.8	Equations with Quadratic Form	182
9.9	Quadratic Inequalities	185

CHAPTER 10 Systems of Equations — 188

10.1	Equations in Two Variables	188
10.2	Independent, Inconsistent, and Dependent Systems of Equations	188
10.3	Solution by Graphing	190
10.4	Analytic Solution	193
10.5	Determinants	196
10.6	Solution by Determinants	197
10.7	Analytic Solution of a System of Equations in Three Variables	201
10.8	Third-Order Determinants	203
10.9	Equations in Three Variables—Solutions by Determinants	206
10.10	The Conics	209
10.11	Systems of One Linear and One Second-Degree Equation	213
10.12	Systems of Two Second-Degree Equations	214

CHAPTER 11 Sequences and Series — 219

11.1	Sequences	219
11.2	Series	221
11.3	Arithmetic Progressions	223
11.4	Arithmetic Series	225
11.5	Geometric Progressions	227
11.6	Geometric Series	229
11.7	Infinite Geometric Series	232
11.8	The Binomial Expansion	235

CHAPTER 12	**Exponential and Logarithmic Functions**	241
12.1	The Exponential Function	241
12.2	The Logarithmic Function	244
12.3	Laws of Logarithms	246
12.4	Common Logarithms	248
12.5	Linear Interpolation	252
12.6	Logarithmic Computation	255
12.7	Logarithmic and Exponential Equations	257

Appendix   261

Answers to Odd-Numbered Problems   265

Index   303

MODERN INTERMEDIATE ALGEBRA

CHAPTER 1

Real Numbers and Their Properties

1.1 ◆ Sets and Symbols

A **set** is a collection of things or objects. Examples of some familiar sets are:
  A set of books.
  A fleet of trucks (another way of saying a set of trucks).
  A navy (another way of saying a set of ships).
  An orchard (another way of saying a set of trees).

The objects or things that make up a set are said to **belong to** the set and are called **members** or **elements** of the set. For example, Monday is a member of the set of days of the week; 3 is a member of the set of odd numbers less than 15.

We must denote a set in such a way that there is no doubt about which elements belong to it. One way to do this is to list all of its members. If we consider the set of natural numbers less than 4, we use the symbol

$$\{1, 2, 3, 4\}$$

to denote this set. When we use this symbol to denote a set the members or elements of the set are enclosed in **braces** { } with commas separating them. The symbol $\{1, 2, 3, 4\}$ is read "the set whose elements are 1, 2, 3, and 4."

We customarily use capital letters as names for sets. For example,

$$A = \{1, 3, 5\}$$

When all the elements of a set are tabulated, as in the set A above, we say that we have **listed** its members.

When there are a great many elements in a set we often abbreviate. For example,

$$S = \{a, b, c, \ldots, x, y, z\}$$

is the set of letters of the English alphabet. The three dots indicate that the letters d through w are also members of the set but have not been listed.

The set of numbers used in counting is called the set of **counting numbers** or the set of **natural numbers,** namely,

$$N = \{1, 2, 3, 4, \ldots\}$$

The set of natural numbers is a nonending set. Notice that to designate a nonending set we again use three dots. Here the three dots mean "and so on in the same manner and continuing indefinitely."

We may designate a set by specifying a characteristic that each element of the set must have. In this case we use the following notation:

$$B = \{x \mid x \text{ is a natural number less than } 5\}$$

We read this symbol "The set of all x such that x is a natural number and x is less than 5." The letter x represents any of the natural numbers 1, 2, 3, or 4. The letter x, when used in this way, is called a **variable** and the set from which we may select a replacement for the variable x is called the **replacement** set, the **domain of the variable** or, simply, the **domain.** The domain in the case of the variable in set B is the set of natural numbers.

The notation used to denote set B above is called **set-builder notation.** Notice that the vertical bar in set-builder notation is read "such that."

When two sets have exactly the same elements, they are said to be **equal.** Thus

$$\{1, 2, 3, 4\} = \{x \mid x \text{ is a natural number and } x \text{ is less than } 5\}$$

and

$$\{x \mid x \text{ is a day of the week}\} = \{\text{Sunday, Monday, Tuesday, Wednesday,} \\ \text{Thursday, Friday, Saturday}\}$$

A set that has no elements is called the **empty set.** We use the symbol ∅ to denote the empty set.

If every member of set B is also a member of set A we say that B is a subset of A. Thus $\{1, 2, 3, 4\}$ is a **subset** of the set of natural numbers, $\{1, 2, 3, 4, \ldots\}$.

A set may contain many elements. For example,

$$P = \{1, 2, 3, 4, \ldots, 9999, 10{,}000\}$$

contains ten thousand elements. If there were any doubt about this we could

1.1 SETS AND SYMBOLS

list all of the elements and count them. A set whose elements can be arranged in some fashion and counted until the last element is reached is called a **finite set**. Set P above is a finite set. It should be noted that a set may have a very large number of elements and still be a finite set. For example, the set of persons who live in China has a fantastically large number of elements but, nevertheless, is a finite set.

A set that is not finite is called an **infinite set**. An example of an infinite set is the set of natural numbers. Since there is no last or greatest natural number, it is impossible to count the elements of the set of natural numbers with the counting coming to an end.

In this book we shall be dealing with sets of numbers. From past experience we are acquainted with the set N of **counting** or **natural numbers**:

$$N = \{1, 2, 3, 4, \ldots\}$$

The set W of **whole numbers** consists of the set of natural numbers and zero:

$$W = \{0, 1, 2, 3, 4, \ldots\}$$

The set I of **integers** consists of the set of whole numbers and their negatives:

$$I = \{\ldots, -3, -2, -1, 0, 1, 2, 3, \ldots\}$$

The set F of **rational numbers** is the set of all numbers that can be represented in the form $\frac{a}{b}$ where a and b are integers and $b \neq 0$. Using set-builder notation we denote the set of rational numbers by

$$F = \left\{ \frac{a}{b} \;\middle|\; a \text{ is an integer}, b \text{ is an integer, and } b \neq 0 \right\}$$

Among the elements of F are $\frac{1}{2}$, $-\frac{1}{4}$, 7, -3, and 0.

Although we denote rational numbers by fractions, it must be emphasized that not all fractions denote rational numbers. For example, the fractions

$$\frac{\sqrt{2}}{3} \quad \text{and} \quad \frac{\pi}{4}$$

are fractions, but they do not denote rational numbers, since $\sqrt{2}$ and π are not integers. The numbers $\sqrt{2}$ and π are elements of a set of numbers called the **irrational numbers**.

We recall from arithmetic that rational numbers may be denoted in decimal form either by a terminating decimal or a repeating, nonterminating decimal. For example,

$$\tfrac{1}{2} = 0.5 \qquad \tfrac{1}{3} = 0.333\ldots$$
$$\tfrac{3}{8} = 0.375 \qquad \tfrac{1}{6} = 0.1666\ldots$$

The set S of **irrational numbers** is the set of all numbers whose decimal representations are nonterminating and nonrepeating. An irrational number *cannot* be represented in the form $\dfrac{a}{b}$ where a and b are integers and $b \neq 0$. Among the elements of S are $\sqrt{2}$, π, $-\sqrt{3}$, and $\sqrt[3]{5}$.

The set R of **real numbers**, which we shall study in this book, consists of the set of rational numbers and the set of irrational numbers.

1.2 ✦ The Equality Relation

We are familiar with many kinds of relations. For example,

> Boston *is east of* Chicago.
> Six *is less than* nine.
> Twelve *is divisible by* two.

"Is east of," "is less than," and "is divisible by" are **relations**. These relations are called binary relations because they relate two elements of a set in a particular order. The most familiar mathematical relation is "equals" or "is equal to."

We use the **equality relation** in all branches of mathematics as well as in other subjects, and in everyday conversation. We now examine the meaning of this relation and state its properties. When we say that "a is equal to b," which we write $a = b$, we mean that a and b are names for the same thing. For example, when we say $1 + 1$ equals 2, we mean that $1 + 1$ and 2 are names for the same number; when we say that five nickels equal one quarter, we mean that five nickels and one quarter name the same amount of money.

The equality relation, symbolized by $=$, has the following properties.

E-1. (*Reflexive Property*) For every a, $a = a$.
E-2. (*Symmetric Property*) For every a and b, if $a = b$, then $b = a$.
E-3. (*Transitive Property*) For every a, b, and c, if $a = b$ and $b = c$, then $a = c$.

Any relation that is reflexive, symmetric and transitive is called an **equivalence relation**.

E-4. (*Substitution Property*) If $a = b$, then a can be replaced by b in any expression containing a.

1.2 THE EQUALITY RELATION

E-5. *(Addition Property)* If $a = b$, then for all c, $a + c = b + c$ and $c + a = c + b$.

E-6. *(Subtraction Property)* If $a = b$, then for all c, $a - c = b - c$ and $c - a = c - b$.

E-7. *(Multiplication Property)* If $a = b$, then for all c, $ac = bc$ and $ca = cb$.

E-8. *(Additive Cancellation Property)* If $a + c = b + c$ or $c + a = c + b$, then $a = b$.

E-9. *(Multiplicative Cancellation Property)* If $ac = bc$ or $ca = cb$ and $c \neq 0$, then $a = b$.

In writing an equality, called an **equation,** we refer to the symbol or symbols to the left of the equal sign as the left member and those to the right of the equal sign as the right member. For example, in the equation

$$2x + 5 = 5 + y$$

$2x + 5$ is the left member and $5 + y$ is the right member.

EXERCISES 1.1

Using braces, list the members of the following sets (Exercises 1–10).

1. The set of the days of the week.
2. The set of the months of the year.
3. The set of the seasons of the year.
4. The set of the first three natural numbers.
5. The set of natural numbers less than 10.
6. The set of counting numbers greater than 8 and less than 12.
7. The set of integers between -5 and 5.
8. The set of the first five odd natural numbers.
9. The set of even natural numbers.
10. The set of odd natural numbers which are less than 20 and divisible by 3.

Use set-builder notation to denote the following sets (Exercises 11–20).

11. The set of taxpayers.
12. The set of soldiers in the United States Army.
13. The set of senators of the Congress of the United States.
14. The set of natural numbers less than 7.
15. The set of integers greater than -3.
16. The set of integers greater than -6 and less than 18.
17. The set of natural numbers greater than 12.
18. The set of even integers.

6   REAL NUMBERS AND THEIR PROPERTIES

19. The set of integers not equal to 4.
20. The set of positive integers.

State whether each of the following sets is infinite or finite (Exercises 21–30).

21. The set of words in the English language.
22. The set of whole numbers.
23. The set of residents of India.
24. The set of integers less than zero.
25. The set of rational numbers between 0 and 1.
26. The set of natural numbers whose numerals end in 0.
27. The set of grains of sand on the beaches in California.
28. The set of integers greater than 9.
29. The set of integral multiples of 3.
30. The set of natural numbers between 6 and 20.

List the members of the following sets (Exercises 31–36).

31. $\{x \mid x \text{ is a natural number and } x \text{ is less than } 4\}$.
32. $\{x \mid x \text{ is a state of the United States whose name begins with the letter O}\}$.
33. $\{x \mid x \text{ is a month of the year whose name begins with } J\}$.
34. $\{x \mid x \text{ is a day of the week whose name begins with } T\}$.
35. $\{x \mid x \text{ is a natural number and } x \text{ is greater than 15 and less than 30}\}$.
36. $\{x \mid x \text{ is an integer and } x \text{ is greater than } -9 \text{ and less than } 9\}$.

Each of the statements below is an application of one of the axioms of the equality relation. Justify each statement by citing the appropriate axiom (Exercises 37–44).

37. If $a = 7$, then $2a = 14$.
38. If $x = 5$, then $5 = x$.
39. If $a = b$ and $a + 8 = 9$, then $b + 8 = 9$.
40. If $c = d$ and $d = e$, then $c = e$.
41. If $x + 3 = 5$, then $x + 1 = 3$.
42. If $2a = c$ and $c = 4e$, then $2a = 4e$.
43. If $8a = 12$, then $2a = 3$.
44. If $z - 6 = 9$, then $z = 15$.

1.3 ◆ The Real Numbers and Their Properties

The set of **real numbers** consists of all **rational numbers** and all **irrational numbers**. Each rational number has a name of the form $\frac{a}{b}$ where $b \neq 0$ and a and b are integers. No irrational number has a name of this form.

All rational numbers can be named by terminating decimals or by repeating, nonterminating decimals. For example,

1.3 THE REAL NUMBERS AND THEIR PROPERTIES

$\frac{1}{2} = 0.5$ $\frac{1}{4} = 0.25$
$-\frac{1}{8} = -0.125$ $\frac{1}{3} = 0.333\ldots$
$\frac{1}{6} = 0.1666\ldots$ $\frac{1}{9} = 0.1111\ldots$
$\frac{3}{2} = 1.5$ $-\frac{7}{3} = -2.3333\ldots$

An irrational number cannot be named in this way. It can, however, be approximated by a decimal. For example, $\sqrt{2} \doteq 1.414$. The symbol $\doteq$ is read "is approximately equal to."

We now list the **properties** or **axioms** of the set, R, of real numbers under the operations of addition and multiplication.

R-1. (*Closure Property of Addition*) For every a and b in R there exists a unique sum, $a + b$, in R.

R-2. (*Commutative Property of Addition*) For every a and b in R, $a + b = b + a$.

R-3. (*Associative Property of Addition*) For every a, b, and c in R, $(a + b) + c = a + (b + c)$.

R-4. (*Identity Element for Addition*) There exists in R an element, 0, called the **additive identity**, such that for every a in R, $a + 0 = 0 + a = a$.

R-5. (*Additive Inverses*) For every element a in R, there exists a unique element, $-a$, in R, called the **additive inverse**, or **opposite**, of a, such that $a + (-a) = 0$. The additive identity, 0, is its own additive inverse.

R-6. (*Closure Property of Multiplication*) For every a and b in R, there exists a unique product, ab, in R.

R-7. (*Commutative Property of Multiplication*) For every a and b in R, $ab = ba$.

R-8. (*Associative Property of Multiplication*) For every a, b, and c in R, $(ab)c = a(bc)$.

R-9. (*Identity Element for Multiplication*) There exists in R an element, 1, called the **multiplicative identity**, such that for every a in R, $a \cdot 1 = 1 \cdot a = a$.

R-10. (*Multiplicative Inverses*) For every element a in R, other than 0, there exists a unique element, $\frac{1}{a}$, in R, called the **multiplicative inverse**, or **reciprocal**, of a, such that $a \cdot \frac{1}{a} = \frac{1}{a} \cdot a = 1$. The multiplicative identity, 1, is its own multiplicative inverse.

R-11. (*Multiplication Property of Zero*) For every a in R, $a \cdot 0 = 0 \cdot a = 0$.

R-12. (*Distributive Property*) For every a, b, and c in R,
$a(b + c) = ab + ac$.

Any set of elements with two operations defined on its elements and which has all the properties listed above is called a **number field** or simply a **field**. The set of real numbers with the operations of addition and multiplication is called the **real number field**.

The axioms or properties of the real number field fall into three classes.

1. Five axioms describing the behavior of the real numbers under addition.
2. Six axioms describing the behavior of the real numbers under multiplication.
3. One axiom connecting addition and multiplication.

Axioms R-1 and R-6 assert that the sum and product of any two real numbers is a real number and that each such sum and product is unique. When the result of an operation on the elements in a set is another element in that set, we say the set is **closed with respect to that operation**. Axioms R-1 and R-6 tell us that the set of real numbers is closed with respect to addition and multiplication.

Axioms R-2 and R-7 state that the order in which we add or multiply real numbers does not affect the sum or product. Thus

$$6 + 7 = 7 + 6 \quad \text{and} \quad 3 \cdot 6 = 6 \cdot 3$$

Axioms R-3 and R-8 state that three terms in a sum or three factors in a product may be associated in either of two ways and that the grouping does not affect the sum or product. Thus since

$$(4 + 2) + 5 = 6 + 5 = 11$$

and

$$4 + (2 + 5) = 4 + 7 = 11$$

it follows that

$$(4 + 2) + 5 = 4 + (2 + 5)$$

Similarly, since

$$(4 \cdot 2) \cdot 5 = 8 \cdot 5 = 40$$

and

$$4 \cdot (2 \cdot 5) = 4 \cdot 10 = 40$$

it follows that

$$(4 \cdot 2) \cdot 5 = 4 \cdot (2 \cdot 5)$$

1.3 THE REAL NUMBERS AND THEIR PROPERTIES

The **generalized associative properties** can be proved to show that the sum of any finite number of terms and the product of any finite number of factors can be indicated unambiguously without parentheses.

As a result of the commutative and associative properties, we may add any number of terms or multiply any number of factors in any order that we please.

Axioms R-4 and R-11 state the basic properties of the number zero. Axiom R-11* can be derived from the other axioms, but we shall accept it without proof. Axiom R-9 states the basic property of the number one. Zero and one are called the **additive** and **multiplicative identities**, respectively.

Axiom R-5 tells us that every real number has an opposite and that the sum of any real number and its opposite is zero. Thus

$$4 + (-4) = 0 \quad \text{and} \quad (-3) + [-(-3)] = 0$$

Since the opposite of any real number a is the unique real number $-a$, it follows that if $a + b = 0$ then $b = -a$. That is, if the sum of two real numbers is zero, each is the opposite, or the additive inverse, of the other. Thus

$$(-3) + [-(-3)] = 0 \quad \text{implies that} \quad -(-3) = 3$$

since we know that

$$3 + (-3) = (-3) + 3 = 0$$

In general, we have, for all a in R,

$$-(-a) = a$$

We observe that the opposite of a positive number is a negative number, that the opposite of a negative number is a positive number, and that 0 is its own opposite.

We now define the **absolute value** of a real number. The absolute value of a positive number is defined to be the number itself; the absolute value of a negative number is its additive inverse—that is, its opposite, which is a positive number; and the absolute value of zero is zero. We write $|x|$ for the absolute value of x. Thus

$$|-5| = -(-5) = 5; \quad |5| = 5; \quad |0| = 0$$

In general,

$$|x| = x \text{ if } x \text{ is nonnegative}$$
$$|x| = -x \text{ if } x \text{ is negative}$$

---

* Axiom R-11 is usually not given as one of the field axioms but is proved as a theorem.

Axiom R-10 tells us that every real number except zero has a **multiplicative inverse** or a **reciprocal** and that the product of any nonzero real number and its multiplicative inverse is 1. Thus

$$2 \cdot \tfrac{1}{2} = 1 \quad \text{and} \quad -4\left(\frac{1}{-4}\right) = 1$$

Axiom R-12 connects the two operations of addition and multiplication. Thus

$$3(4 + 5) = (3 \cdot 4) + (3 \cdot 5)$$

that is, $3 \cdot 9 = 12 + 15$, which is true since $27 = 27$. Using R-12 and R-3, we can show that

$$a(b + c + d) = ab + ac + ad$$
$$a(b + c + d + e) = ab + ac + ad + ae$$

and so forth. This is called the **generalized distributive property**. Axiom R-7 assures us that

$$a(b + c + d + e) = (b + c + d + e)a$$

We now define the order relation between pairs of real numbers. The concept of order can be defined in terms of addition. The order relation "is less than" is denoted by the symbol $<$ and is defined as follows: **if a and b are real numbers, then $a < b$** (or $b > a$, read: "b is greater than a") **if and only if there exists a positive real number r such that $a + r = b$**. Thus

$$-2 < 4 \quad \text{because } (-2) + 6 = 4 \text{ and 6 is positive}$$
$$-8 < -5 \quad \text{because } (-8) + 3 = -5 \text{ and 3 is positive}$$
$$2 < 9 \quad \text{because } 2 + 7 = 9 \text{ and 7 is positive}$$

The order relation has the following properties.

0-1. (*Trichotomy Property*) If a and b are in R, then one and only one of the following is true:

$$a < b; \; a = b; \; a > b$$

0-2. (*Transitive Property*) If a, b, and c are elements of R and $a < b$ and $b < c$, then $a < c$.

0-3. (*Addition Property of Inequality*) If a, b, and c are elements of R and $a < b$, then $a + c < b + c$.

0-4. (*Multiplication Property of Inequality for Positive Numbers*) If a, b, and c are elements of R, $a < b$, and c is positive, then $ac < bc$.

1.3 THE REAL NUMBERS AND THEIR PROPERTIES

0-5. (*Multiplication Property of Inequality for Negative Numbers*) If a, b, and c are elements of R, $a < b$, and c is negative, then $ac > bc$.

The order relation may be stated by using the following symbols:

$<$ read "is less than"
$>$ read "is greater than"
$\leq$ read "is less than or equal to"
$\geq$ read "is greater than or equal to"

EXERCISES 1.2

Each of the statements below is an application of one of the axioms of the real numbers. Justify each statement by citing the appropriate axiom (Exercises 1–21).

1. $(3)(-2)$ is a real number.
2. $3 \cdot (8k) = (3 \cdot 8)k$.
3. $2 + \frac{1}{2}$ is a real number.
4. $3 + 4 = 4 + 3$.
5. $8(4 + 3) = 8 \cdot 4 + 8 \cdot 3$.
6. $3 + (-3) = 0$.
7. $(\frac{1}{2})(2) = 1$.
8. $3 \cdot 2 + 5 \cdot 2 = (3 + 5)2$.
9. $(-9) + 0 = -9$.
10. $-(-3) + (-3) = 0$.
11. $(\sqrt{2} + 6) + 0 = \sqrt{2} + 6$.
12. $2(4 \cdot 7) = (2 \cdot 4)7$.
13. $3 + (6 + \sqrt{5}) = (3 + 6) + \sqrt{5}$.
14. $-2 + [-(-2)] = 0$.
15. $2 \cdot (4 + 6) = (4 + 6) \cdot 2$.
16. $8 + (3 + c) = (8 + 3) + c$.
17. $a(b + c) = (b + c)a$.
18. $(3)(\frac{1}{3}) = 1$.
19. $\left(\dfrac{1}{-4}\right)(-4) = 1$.
20. $2(3 + 4 + 7) = 2 \cdot 3 + 2 \cdot 4 + 2 \cdot 7$.
21. $\sqrt{5}[\sqrt{7} + (-\sqrt{7})] = \sqrt{5} \cdot 0$.

Which of the following statements are true (Exercises 22–34)?

22. If a is a positive real number then $-a$ is negative.
23. If $a = 0$, then $|a| = 0$.
24. If a is a negative real number, then $-a$ is negative.

25. If a is a negative real number, then $-a$ is positive.
26. If $-a$ is a positive real number, then a is negative.
27. $|-2| = |2|$.
28. $|-7| = -7$.
29. $|-100| < |100|$.
30. $0 \neq |0|$.
31. $-1 < |-19|$.
32. $|-4| = |4|$.

33. If x and y are real numbers and $x > y$, then $|x| > |y|$.
34. If x and y are real numbers and $x \neq y$, then $|x| \neq |y|$.
35. Write the multiplicative inverse of each of the following.
    (a) 7
    (b) $-\frac{1}{2}$
    (c) x, $x \neq 0$
    (d) $a + b$, $a \neq -b$
    (e) $x^2 - t^2$, $x^2 \neq t^2$
    (f) $\frac{3x}{y}$, $x \neq 0$, $y \neq 0$
    (g) $5x - 2y$, $5x \neq 2y$
    (h) $\sqrt{5}$

36. Write the additive inverse of each of the following.
    (a) 6
    (b) -4
    (c) $-\frac{1}{4}$
    (d) 7
    (e) $x + y$
    (f) $x - y$
    (g) x^2
    (h) $-(-3)$

Replace "*" with =, <, or >, to make true statements (Exercises 37–51).

37. $-8 * 18$
38. $4 * 2 + \sqrt{2}$
39. $-15 * 15$
40. $-14 * -18$
41. $-\frac{1}{2} * -\frac{1}{4}$
42. $|-3| * |-\frac{7}{2}|$
43. $16 * -17$
44. $\frac{3}{4} * \frac{7}{8}$
45. $|5| + 1 * 5$
46. $|-5| * |5|$
47. $|-(-3)| * |-3|$
48. $|7| + |-7| * 0$
49. $-3 * |-10|$
50. $-[-(-5)] * 5$
51. $-(-2) * |2|$

52. Which of the following statements are true for *all* real number values of the variables involved? Assume that there are no divisions by zero.
    (a) $-(-b) = b$
    (b) $-(-(-a)) = a$
    (c) $|a| = |-a|$
    (d) $-a = |-a|$
    (e) $-(-a) = |a|$
    (f) $a < |-a|$
    (g) $|a| + |b| = |a + b|$
    (h) $a + b = |a| + |b|$
    (i) $\frac{1}{a} + \frac{1}{b} = \frac{1}{a+b}$
    (j) $\frac{1}{a} \cdot \frac{1}{b} = \frac{1}{ab}$
    (k) $[a \cdot 1] \cdot \frac{1}{a} = a \cdot \frac{1}{a}$
    (l) $|y| = -y$

1.4 ◆ Prime and Composite Numbers

In this section we shall discuss only the natural numbers: 1, 2, 3, ... In the product abc, the numbers a, b, and c are called **factors.** Thus 2 and 5 are factors of 10 because $10 = 2 \cdot 5$ and 1 and 17 are factors of 17, since $17 = 1 \cdot 17$. A natural number greater than or equal to 2 is either a **prime** number or a **composite** number. A **prime number,** or simply a **prime,** is a natural number greater than 1 which has only two factors: 1 and the number itself. The first few primes are

$$2, 3, 5, 7, 11, 13, 17, 19, 23$$

A natural number, n, greater than 1, which has at least one factor greater than 1 and less than n, is called a **composite** number. The first few composite numbers are

$$4, 6, 8, 9, 10, 12, 14, 15, 16$$

When a composite number is written as the product of prime factors only, we say that it is **factored completely.** The natural numbers below are factored completely:

$$15 = 3 \cdot 5$$
$$16 = 2 \cdot 2 \cdot 2 \cdot 2 = 2^4$$
$$28 = 2 \cdot 2 \cdot 7 = 2^2 \cdot 7$$
$$36 = 2 \cdot 2 \cdot 3 \cdot 3 = 2^2 \cdot 3^2$$

The *Fundamental Theorem of Arithmetic* assures us that *every composite natural number can be written as a product of primes in one and only one way, except possibly for the order in which the prime factors are written.* We shall accept the Fundamental Theorem of Arithmetic without proof.

We use the terms "prime" and "composite" in referring to natural numbers only. Integers, rational numbers, and real numbers that are not natural numbers are not defined as either prime or composite.

A negative integer may always be written as the product of -1 and a natural number. For example, $-32 = (-1)(32)$ and $-16 = (-1)(16)$.

A natural number can be written as the product of two natural numbers in only a finite number of ways. For example,

$$48 = 1 \cdot 48 = 2 \cdot 24 = 3 \cdot 16 = 4 \cdot 12 = 6 \cdot 8$$

On the other hand, if real numbers are considered as factors, natural numbers can be expressed as the product of two factors in an infinite number of ways. For example,

$$48 = (\tfrac{1}{2})(96) = (-\tfrac{1}{2})(-96) = (\tfrac{1}{6})(288) = (\tfrac{3}{4})(\tfrac{192}{3})$$

and so forth.

EXERCISES 1.3

1. List the even primes.
2. Find all the pairs of natural number factors of the following.
   (a) 30
   (b) 24
   (c) 64
   (d) 81
   (e) 144
   (f) 205

Express each of the following integers as the product of -1 and a natural number (Exercises 3–8).

3. -45
4. -101
5. -34
6. -98
7. -516
8. -69

Express each of the following natural numbers in completely factored form (Exercises 9–20).

9. 72
10. 56
11. 125
12. 106
13. 49
14. 144
15. 60
16. 48
17. 1000
18. 5246
19. 299
20. 421

1.5 ◆ Addition and Subtraction of Real Numbers

Real numbers are of little use unless we know how to add, subtract, multiply, and divide them. We shall now develop rules for the addition and subtraction of real numbers. From R-1 we know that the sum, $a + b$, of two real numbers, a and b, is a unique real number. We define the **difference**, $a - b$, of two real numbers, a and b, to be the unique real number c such that $b + c = a$.

By previous experience we know how to find the sum, $a + b$, of two *nonnegative* real numbers, a and b. We shall now show that if a and b are nonnegative real numbers then

$$(-a) + (-b) = -(a + b)$$

that is, we shall develop a rule for adding two *negative* real numbers. We know that

$$a + (-a) = 0$$

1.5 ADDITION AND SUBTRACTION OF REAL NUMBERS

and
$$b + (-b) = 0$$
by the additive inverse axiom (R-5). By the addition property of equality and the additive identity axiom we have
$$[a + (-a)] + [b + (-b)] = 0 + 0 = 0$$
Using the associative and commutative properties of addition, we have:

$\{[a + (-a)] + b\} + (-b) = 0$ associative property
$\{a + [(-a) + b]\} + (-b) = 0$ associative property
$\{a + [b + (-a)]\} + (-b) = 0$ commutative property
$[(a + b) + (-a)] + (-b) = 0$ associative property
$(a + b) + [(-a) + (-b)] = 0$ associative property

Since $(a + b) + [(-a) + (-b)] = 0$ as shown above, and since $(a + b) + [-(a + b)] = 0$ (because any element plus its additive inverse is zero), we conclude that
$$(-a) + (-b) = -(a + b)$$
because of the uniqueness of additive inverses.

We have shown that we can add two negative numbers by taking the opposite of the sum of their additive inverses. This is the same as the opposite of the sum of their absolute values. We express this rule symbolically by saying that, **whenever p and q are negative, then**
$$p + q = -(|p| + |q|)$$

Now let us consider how we add two real numbers when one is negative and the other is nonnegative. Suppose a and b are positive real numbers and we want to find $a + (-b)$. We must consider two cases: (1) $a > b$ and (2) $a < b$.

If $a > b$ there is a positive number c for which $b + c = a$. Now,

$a + (-b) = (b + c) + (-b)$ substitution property of equality
$ = (c + b) + (-b)$ commutative property of addition
$ = c + [b + (-b)]$ associative property of addition
$ = c + 0$ additive inverse axiom
$ = c$ additive identity axiom

We defined the difference, $a - b$, of two real numbers, a and b, to be the real number c such that $b + c = a$. Hence c is the difference of a and b and, therefore, $a + (-b) = a - b$ by the substitution property. We have found that **if p is a positive number and q is a negative number and if also $|p| > |q|$, then**

$$p + q = |p| - |q|$$

Now let us find the sum $a + (-b)$ in the case in which a and b are positive numbers and $a < b$. Since $a < b$, we can find a positive real number c such that

$$a + c = b \quad \text{or} \quad c = b - a$$

Then

$$\begin{aligned}
a + (-b) &= a + [-(a + c)] & & \text{substitution property of equality} \\
&= a + [(-a) + (-c)] & & \text{the rule for addition of two negative numbers} \\
&= [a + (-a)] + (-c) & & \text{association property of addition} \\
&= 0 + (-c) & & \text{additive inverse axiom} \\
&= -c & & \text{additive identity axiom}
\end{aligned}$$

But $c = b - a$, so we have

$$a + (-b) = -(b - a)$$

We have thus found that **if p is a positive number and q is a negative number and if also $|p| < |q|$, then**

$$p + q = -(|q| - |p|)$$

From the above we see that if a and b are positive, then

$$a + (-b) = a - b \quad \text{whenever} \quad a > b$$

and

$$a + (-b) = -(b - a) \quad \text{whenever} \quad a < b$$

We may now formulate the rules for the addition of real numbers.

1. If $p \geq 0$ and $q \geq 0$, then $p + q = |p| + |q|$. Thus

$$3 + 5 = |3| + |5| = 8$$

2. If $p < 0$ and $q < 0$, then $p + q = -(|p| + |q|)$. Thus

$$(-3) + (-5) = -(|-3| + |-5|) = -(3 + 5) = -8$$

3. If $p \geq 0$ and $q < 0$ and $|p| > |q|$, then $p + q = |p| - |q|$. Thus

$$5 + (-3) = |5| - |-3| = 5 - 3 = 2$$

1.5 ADDITION AND SUBTRACTION OF REAL NUMBERS

4. If $p \geq 0$ and $q < 0$ and $|p| < |q|$, then $p + q = -(|q| - |p|)$. Thus
$$3 + (-5) = -(|-5| - |3|) = -(5 - 3) = -2$$

We defined the difference, $a - b$, of two real numbers, a and b, to be the real number c such that $b + c = a$. Applying the addition property of equality with the number $-b$, we have

$$\begin{aligned}
a + (-b) &= (b + c) + (-b) \\
&= (c + b) + (-b) & \text{commutative property of addition} \\
&= c + [b + (-b)] & \text{associative property of addition} \\
&= c + 0 & \text{additive inverse axiom} \\
&= c & \text{additive identity axiom}
\end{aligned}$$

But $c = a - b$ and hence
$$a - b = a + (-b)$$

That is, *subtracting the real number b from the real number a is the same as adding* $-b$, *the opposite of b, to a.*

EXERCISES 1.4

Perform the indicated operations (Exercises 1–24).

1. $2 + 4$
2. $3 + (-4)$
3. $(-5) + (-8)$
4. $8 + (-3)$
5. $(-7) + (-6)$
6. $(-8) - (-4)$
7. $(-9) - (-4)$
8. $7 - 6$
9. $-2 - (-7)$
10. $19 + (-28)$
11. $-105 - (-108)$
12. $86 + (-137)$
13. $(-4) + (-8) + 12$
14. $(-8) + (-8) + (-3)$
15. $-8 - 24 + (-13)$
16. $8 - 6 + 8$
17. $-42 + (-3) + 36$
18. $26 + (-84) + 42$
19. $17 + (-18) - 37$
20. $26 + (-18) - 17$
21. $-13 - (-8) + (-72)$
22. $26 + (-17) + (-23)$
23. $-26 + (-17) + (-23)$
24. $17 + (-14) - 17 + (-18)$

25. Express each of the following as the sum of two real numbers.
    (a) $6 - (-7)$
    (b) $8 - 15$
    (c) $-12 - (-4)$
    (d) $-16 - 8$
    (e) $-20 - (-18)$
    (f) $72 - 36$

26. If a and b are real numbers and $a > b > 0$, is $a + (-b)$ positive or negative?
27. If a and b are real numbers and $a > b > 0$, is $b + (-a)$ positive or negative?
28. If a and b are real numbers and $0 > a > b$, is $a + (-b)$ positive or negative?

REAL NUMBERS AND THEIR PROPERTIES

29. If a and b are real numbers and $0 > a > b$, is $b + (-a)$ positive or negative?
30. A man puts $250 in his checking account when it is opened. During the month he wrote checks for $80, $60, $100, and $50. The bank sent him a letter telling him that his account was overdrawn. How much was the account short?
31. The temperature at 6:00 A.M. is 8° and rises 15° by 10:00 A.M. What is the temperature at 10:00 A.M.?
32. The highest recorded temperature at Detroit, Michigan is 104°. The coldest recorded temperature is $-24°$. What is the difference in temperature between the coldest and the hottest?
33. The average altitudes, in feet, of some of the oceans and seas are given below:

Pacific Ocean	$-14{,}048$ ft
Atlantic Ocean	$-12{,}800$ ft
Arctic Ocean	$-3{,}953$ ft
Indian Ocean	$-13{,}002$ ft
North Sea	-308 ft
English Channel	-190 ft
Persian Gulf	-82 ft

  (a) What is the difference in altitude between the Pacific Ocean and the Persian Gulf?
  (b) What is the difference in altitude between the North Sea and the Atlantic Ocean?
  (c) What is the difference in altitude between the English Channel and the Indian Ocean?

34. The temperature at 6:00 A.M. was 42°, and it was 70° outside. By 1:00 P.M. the temperature had risen 26° outdoors and had fallen 5° indoors. What is the difference between the outdoor and indoor temperatures at 1:00 P.M.?
35. With what real numbers would you replace the variables in the sentences below to make true statements?

  (a) $7 - x = 4$
  (b) $\frac{1}{2} + k = 0$
  (c) $-\frac{2}{3} + m = \frac{1}{6}$
  (d) $v - \frac{3}{5} = 1$
  (e) $\frac{2}{3} - k = \frac{5}{3}$
  (f) $t + (-3) = 7$
  (g) $m - (-5) = -16$
  (h) $\frac{2}{3} - (-\frac{1}{3}) = p$
  (i) $\frac{1}{7} + r = -\frac{4}{21}$
  (j) $s - (-\frac{3}{5}) = 1$

Insert one of the symbols, $=$, $<$, $>$, between the pairs of expressions to give true statements (Exercises 36–45).

36. $|2 + 3|$, $|2| + |3|$.
37. $|(-7) + (-2)|$, $|-7| + |-2|$.
38. $|125 + (-125)|$, $|125| + |-125|$.

1.6 MULTIPLICATION AND DIVISION OF REAL NUMBERS

39. $|-4 + 16|, |-4| + |16|$.
40. $|-\frac{1}{2} + \frac{1}{2}|, |-\frac{1}{2}| + |\frac{1}{2}|$.
41. $|(-4) + (-3)|, |-4| + |-3|$.
42. $|-2 + (-2)|, |-2| + |-2|$.
43. $|-\frac{3}{5} - (-\frac{1}{5})|, |-\frac{3}{5}| - |-\frac{1}{5}|$.
44. $|\frac{1}{4}| + |\frac{1}{2}|, |-\frac{1}{4}| + |-\frac{1}{2}|$.
45. $|\frac{1}{4} + (-\frac{2}{3})|, |\frac{1}{4}| + |-\frac{2}{3}|$.

1.6 ♦ Multiplication and Division of Real Numbers

The product of two positive real numbers is a positive real number. That is, if a and b are positive real numbers, then ab is also a positive real number. We need to develop rules for multiplication of a positive and a negative real number and for multiplication of two negative real numbers. We begin with the case of a product, pq, where p is positive and q is negative.

Let us now consider the real numbers a and $-b$, where a and b are both positive. We know that

$$a + (-b) = 0 \quad \text{additive inverse axiom}$$
$$a[b + (-b)] = 0 \quad \text{multiplication properties of zero and equality}$$
$$ab + a(-b) = 0 \quad \text{distributive property}$$

But we also know that

$$ab + [-(ab)] = 0$$

by the additive inverse axiom, so that the uniqueness of the additive inverse tells us that

$$a(-b) = -(ab)$$

We can conclude that **if p is a positive real number and $(-q)$ is a negative real number, then**

$$p(-q) = -(|p||q|)$$

The commutative property of multiplication allows us to multiply a positive number and a negative number, regardless of which of them is expressed first, hence $p(-q) = (-q)p = -pq$.

We now determine the rule for multiplying two negative numbers. We consider the real numbers $-a$ and $-b$, where a and b are positive real numbers. We have

$$b + (-b) = 0 \quad \text{additive inverse axiom}$$
$$(-a)[b + (-b)] = 0 \quad \text{multiplication properties of zero and equality}$$
$$(-a)b + (-a)(-b) = 0 \quad \text{distributive property}$$

But $(-a)b$ is the product of a negative number and a positive number, and we have just shown that, when a and b are positive, $(-a)b = -(ab)$ so that

$$-(ab) + (-a)(-b) = 0 \quad \text{substitution property of equality}$$

Thus

$$(-a)(-b) = -(-(ab)) \quad \text{additive inverse axiom}$$

But we have already determined that the opposite of the opposite of a real number is the real number itself, so we conclude that

$$(-a)(-b) = ab$$

Thus, if p and q are negative real numbers, then

$$pq = |p| \cdot |q|$$

We summarize by stating the following rules.

If $p \geq 0$ and $q \geq 0$, then $pq = |p||q|$.
If $p \geq 0$ and $q < 0$, then $pq = -(|p||q|)$.
If $p < 0$ and $q \geq 0$, then $pq = -(|p||q|)$.
If $p < 0$ and $q < 0$, then $pq = |p||q|$.

If a and b are real numbers and $b \neq 0$, we define the **quotient** $\frac{a}{b}$ to be the real number c such that $a = bc$. In the equation $\frac{a}{b} = c$ we call a the **dividend**, b the **divisor**, and c the **quotient**. The divisor cannot be 0 since $\frac{a}{0} = c$ would mean $c(0) = a$ and $a = 0$ by the multiplication property of 0. But then c could be any number and we need a definition of an arithmetic operation which produces a unique result. Thus *division by zero is not permitted* in this system.

Since $\frac{a}{b} = c$ means that $bc = a$ the sign of the quotient of two real numbers can be determined by the rules for the sign of the product of two real numbers.

If p and q are positive and $\frac{p}{q} = r$ then $p = qr$. Since p is positive, the product qr is positive. But, since q is positive, r must also be positive; since otherwise the product qr would be negative. Thus the quotient of two positive real numbers is a positive real number.

1.6 MULTIPLICATION AND DIVISION OF REAL NUMBERS

If p is positive and q is negative and $\frac{p}{q} = r$, then $p = qr$. Since p is positive, the product qr is positive. But, since q is negative, r must also be negative; otherwise the product qr would be negative. Thus a quotient with a positive dividend and a negative divisor must be negative.

If p is negative and q is positive and $\frac{p}{q} = r$, then $p = qr$. Since p is negative, the product qr is negative. But, since q is positive, r must be negative; otherwise the product qr would be positive. Thus a quotient with a negative dividend and a positive divisor must be negative.

Finally, if p and q are negative and $\frac{p}{q} = r$, then $p = qr$. Since p is negative, the product qr is also negative. But, since q is negative, r must be positive; otherwise the product qr would be positive. Thus the quotient of two negative real numbers is a positive real number.

We summarize the rules for dividing real numbers by:

NEVER DIVIDE BY ZERO.

If $p > 0$ and $q > 0$, then $\frac{p}{q} = \frac{|p|}{|q|}$.

If $p > 0$ and $q < 0$, then $\frac{p}{q} = -\frac{|p|}{|q|}$.

If $p < 0$ and $q > 0$, then $\frac{p}{q} = -\frac{|p|}{|q|}$.

If $p < 0$ and $q < 0$, then $\frac{p}{q} = \frac{|p|}{|q|}$.

EXERCISES 1.5

Find the products (Exercises 1–21).

1. $2 \cdot 7$
2. $4(-3)$
3. $(-5)(-9)$
4. $9(-5)$
5. $(-9)(-6)$
6. $(18)(-5)$
7. $(12)(-2)$
8. $(-7)(-5)$
9. $(-1)(3)$
10. $(-1)(-3)$
11. $(-8)(4)$
12. $(16)(-3)$
13. $15 \cdot 4$
14. $(-9)(-8)$
15. $(-1)(-3)(-7)$
16. $(3)(-4)(-5)$
17. $(-3)(-2)(2)$
18. $(-2)(-8)(-10)$
19. $3 \cdot 2 \cdot 0$
20. $-4 \cdot 3 \cdot 0$
21. $(-1)(6)(5)$

REAL NUMBERS AND THEIR PROPERTIES

Find the quotients (Exercises 22–36).

22. $\dfrac{25}{5}$ 27. $\dfrac{-125}{5}$ 32. $\dfrac{-88}{-11}$

23. $\dfrac{-6}{-2}$ 28. $\dfrac{81}{9}$ 33. $\dfrac{-342}{9}$

24. $\dfrac{3}{-1}$ 29. $\dfrac{0}{54}$ 34. $\dfrac{128}{-4}$

25. $\dfrac{40}{10}$ 30. $\dfrac{336}{-3}$ 35. $\dfrac{-7}{-1}$

26. $\dfrac{-16}{8}$ 31. $\dfrac{144}{-8}$ 36. $\dfrac{84}{12}$

37. Fill in the blanks to make true statements.

   (a) $\dfrac{-18}{2} = -9$ because $(2)\underline{\quad} = -18$.

   (b) $\dfrac{125}{25} = 5$ because $(25)\underline{\quad} = 125$.

   (c) $\dfrac{-100}{-10} = 10$ because $(-10)\underline{\quad} = -100$.

   (d) $\dfrac{36}{-4} = -9$ because $(-4)\underline{\quad} = 36$.

Write the following products as quotients (Exercises 38–43).

38. $5(\tfrac{1}{2})$ 40. $12(\tfrac{1}{3})$ 42. $7(\tfrac{1}{4})$
39. $8(\tfrac{1}{4})$ 41. $15(\tfrac{1}{8})$ 43. $18(\tfrac{1}{16})$

44. Which of the following statements are true? All of the variables represent real numbers.

   (a) If $x < 0$ and $y > 0$, then $xy < 0$.
   (b) If $x > 0$ and $y > 0$, then $xy < 0$.
   (c) If $x < 0$ and $y > 0$, then $xy > 0$.
   (d) If $x > 0$ and $y < 0$, then $xy < 0$.
   (e) $\dfrac{a}{b} = a\left(\dfrac{1}{b}\right)$, $b \neq 0$.
   (f) $(-1)x = -x$.
   (g) $(-x)(-1) = -x$.
   (h) $x \cdot 0 = 0$.

What real number replacements of the variables make true statements (Exercises 45–56)?

45. $2x = 16$ 47. $5s = -25$
46. $-16x = 8$ 48. $2t = 5$

49. $(-4)(\frac{3}{2}) = m$
50. $z(-\frac{1}{2}) = \frac{1}{6}$
51. $\frac{3}{4}y = \frac{1}{2}$
52. $(-\frac{3}{5})k = \frac{3}{10}$
53. $(-\frac{5}{8})p = \frac{3}{7}$
54. $(\frac{5}{2})(-\frac{3}{4}) = y$
55. $\frac{2}{3}y = \frac{11}{7}$
56. $\frac{5}{8}x = \frac{3}{5}$

Insert one of the symbols, $=, <, >$, between each pair of expressions to make true statements (Exercises 57–60).

57. $|2 \cdot 3|, |2| \cdot |3|$
58. $|\frac{1}{2}(-\frac{3}{4})|, |\frac{1}{2}| \cdot |\frac{3}{4}|$
59. $|0 \cdot 0|, |0| \cdot |0|$
60. $|(-3)(-7)|, |-3| \cdot |-7|$

CHAPTER 1   REVIEW

1. Using braces, list the members of the following sets.
   (a) $\{x \mid x$ is an integer and x is greater than $6\}$.
   (b) $\{x \mid x$ is an odd natural number$\}$.
   (c) $\{x \mid x$ is a natural number and x is greater than 2 and less than $12\}$.
   (d) $\{x \mid x$ is an integer less than $-2\}$.

2. Each of the statements below is an application of one of the axioms of the equality relation. Justify each statement by citing the appropriate axiom.
   (a) If $x = 3$, then $x + 2 = 5$.
   (b) If $p = q$ and $q = r$, then $p = r$.
   (c) If $x - 7 = 14$, then $x = 21$.

3. Justify each statement below by citing the appropriate axiom of the real number system.
   (a) $\frac{1}{4} + \frac{1}{2} = \frac{1}{2} + \frac{1}{4}$.
   (b) $\sqrt{2}\left(\frac{1}{\sqrt{2}}\right) = 1$.
   (c) $\sqrt{3}(\sqrt{5} + \sqrt{2}) = \sqrt{3} \cdot \sqrt{5} + \sqrt{3} \cdot \sqrt{2}$.
   (d) $-3 + [-(-3)] = 0$.

4. Factor each of the following natural numbers completely.
   (a) 48   (b) 72   (c) 144   (d) 1000

Perform the indicated operations (Exercises 5–12).

5. $(-3) + 4 - (-6)$
6. $14 - (-2) + (-7)$
7. $(-3)(4)(-2)$
8. $\frac{-125}{5}$
9. $\frac{342}{-3}$
10. $(2)(-7)(-3)(-2)$
11. $\frac{-36}{-6}$
12. $(18) \div (-2) - 16 + (-3)$

13. What is the multiplicative inverse of $\sqrt{3}$?
14. What is the additive inverse of $\sqrt{2}$?
15. What is the multiplicative identity of the set of real numbers?
16. What is the additive identity of the set of real numbers?

Which of the following statements are true for all real number values of the variable? Assume that no divisors are zero (Exercises 17–20).

17. $|a| = |-a|$ 19. $a + (-a) = 2a$

18. $a \cdot \dfrac{1}{a} = 1$ 20. $-(-a) = a$

CHAPTER 2

Exponents and Radicals

2.1 ◆ Laws of Positive Integer Exponents

We define $2 \cdot 2 \cdot 2$ as 2^3, where 2 is called the **base,** 3 is called the **exponent,** and 2^3 is called the **power.** The exponent, when it is a positive integer, tells how many times the base is used as a factor. Thus

$$3 \cdot 3 = 3^2$$
$$6 \cdot 6 \cdot 6 \cdot 6 \cdot 6 = 6^5$$
$$10 \cdot 10 \cdot 10 \cdot 10 = 10^4$$

In general, when n is a positive integer

$$a^n = \underbrace{a \cdot a \cdot a \cdots a}_{n \text{ factors}}$$

The real number a is called the **base,** the positive integer n is called the **exponent,** and a^n is called the **nth power** of a. We form this definition by saying that a positive integer exponent tells us how many times the base is used as a factor of the product which is the nth power of the real number a.

This definition also says that, for any real number, a,

$$a^1 = a$$

The following laws are called the **Laws of Exponents.**

LAW 1. If m and n are positive integers, and a is any real number, then

$$a^m a^n = a^{m+n}$$

PROOF

$$a^m a^n = \underbrace{(a \cdot a \cdots a)}_{m \text{ factors}} \underbrace{(a \cdot a \cdots a)}_{n \text{ factors}}$$

$$= \underbrace{a \cdot a \cdots a}_{m + n \text{ factors}}$$

$$= a^{m+n}$$

EXAMPLES

$$x^3 \cdot x^5 = x^{3+5} = x^8$$
$$y^5 \cdot y^9 = y^{5+9} = y^{14}$$
$$x^a \cdot x^{2a} = x^{3a} \text{ (where } a \text{ is a positive integer)}$$
$$(y - x)^3 (y - x)^2 = (y - x)^5$$

LAW 2. If m and n are positive integers, $m > n$, and $a \neq 0$ is a real number, then

$$\frac{a^m}{a^n} = a^{m-n}$$

PROOF

$$\frac{a^m}{a^n} = \frac{\overbrace{a \cdot a \cdots a}^{m \text{ factors}}}{\underbrace{a \cdot a \cdots a}_{n \text{ factors}}}$$

$$= \frac{\overbrace{(a \cdot a \cdots a)}^{n \text{ factors}} \overbrace{(a \cdot a \cdots a)}^{m - n \text{ factors}}}{\underbrace{a \cdot a \cdots a}_{n \text{ factors}}}$$

$$= \underbrace{(a \cdot a \cdots a)}_{m - n \text{ factors}} \cdot 1$$

$$= a^{m-n}$$

EXAMPLES

$$\frac{x^5}{x^3} = x^{5-3} = x^2, \; x \neq 0$$

$$\frac{y^{3m}}{y^{2m}} = y^m, \text{ where } m \text{ is a positive integer, } y \neq 0$$

$$\frac{(x - y)^8}{(x - y)^2} = (x - y)^6, \; x \neq y$$

2.1 LAWS OF POSITIVE INTEGER EXPONENTS

LAW 3. If m and n are positive integers, $m < n$, and $a \neq 0$ is a real number, then

$$\frac{a^m}{a^n} = \frac{1}{a^{n-m}}$$

PROOF

$$\frac{a^m}{a^n} = \frac{\overbrace{a \cdot a \cdots a}^{m \text{ factors}}}{\underbrace{a \cdot a \cdots a}_{n \text{ factors}}}$$

$$= \frac{\overbrace{a \cdot a \cdots a}^{m \text{ factors}}}{\underbrace{(a \cdot a \cdots a)}_{m \text{ factors}} \underbrace{(a \cdot a \cdots a)}_{n-m \text{ factors}}}$$

$$= \frac{1}{\underbrace{a \cdot a \cdots a}_{n-m \text{ factors}}}$$

$$= \frac{1}{a^{n-m}}$$

EXAMPLES

$$\frac{x^3}{x^5} = \frac{1}{x^{5-3}} = \frac{1}{x^2}, \quad x \neq 0$$

$$\frac{y^{2k}}{y^{5k}} = \frac{1}{y^{3k}}, \quad \text{where } k \text{ is a positive integer, } y \neq 0$$

$$\frac{(x-y)^5}{(x-y)^7} = \frac{1}{(x-y)^2}, \quad x \neq y$$

LAW 4. If m is a positive integer and a and b are any real numbers, then

$$(ab)^m = a^m b^m$$

PROOF

$$(ab)^m = (ab)(ab) \cdots (ab) \qquad \text{with } m \text{ factors}$$
$$= [\underbrace{(a)(a) \cdots (a)}_{m \text{ factors}}][\underbrace{(b)(b) \cdots (b)}_{m \text{ factors}}] \qquad \text{by the associative and commutative laws of multiplication}$$
$$= a^m b^m$$

EXAMPLES

$$(2x)^3 = 2^3 \cdot x^3 = 8x^3$$
$$(3xy)^4 = 3^4 \cdot x^4 \cdot y^4 = 81x^4y^4$$

LAW 5. If m is a positive integer, a and b are real numbers, and $b \neq 0$, then

$$\left(\frac{a}{b}\right)^m = \frac{a^m}{b^m}$$

PROOF

$$\left(\frac{a}{b}\right)^m = \underbrace{\left(\frac{a}{b}\right)\left(\frac{a}{b}\right) \cdots \left(\frac{a}{b}\right)}_{m \text{ factors}} \quad \text{with } m \text{ factors}$$

$$= \frac{\overbrace{a \cdot a \cdot a \cdots a}^{m \text{ factors}}}{\underbrace{b \cdot b \cdot b \cdots b}_{m \text{ factors}}}$$

$$= \frac{a^m}{b^m}$$

EXAMPLES

$$\left(\frac{x}{y}\right)^3 = \frac{x^3}{y^3}, \, y \neq 0$$

$$\left(\frac{2x}{3y}\right)^4 = \frac{(2x)^4}{(3y)^4} = \frac{2^4 x^4}{3^4 y^4} = \frac{16x^4}{81y^4}, \, y \neq 0$$

$$\left(\frac{x-y}{x+y}\right)^3 = \frac{(x-y)^3}{(x+y)^3}, \, (x+y) \neq 0$$

LAW 6. If m and n are positive integers and a is any real number, then

$$(a^m)^n = a^{mn}$$

PROOF

$$(a^m)^n = (a^m)(a^m) \cdots (a^m) \quad \text{with } n \text{ factors}$$
$$= a^{m+m+\ldots+m} \quad \text{with } n \text{ terms in the exponent}$$
$$= a^{mn}$$

EXAMPLES

$$(2x^2)^3 = 2^3(x^2)^3 = 2^3 \cdot x^6 = 8x^6$$
$$(x^2y^3)^5 = (x^2)^5(y^3)^5 = x^{10}y^{15}$$

2.1 LAWS OF POSITIVE INTEGER EXPONENTS

We restate these six laws of positive integer exponents below.

If m and n are positive integers and a and b are any real numbers:

1. $a^m \cdot a^n = a^{m+n}$
2. $\dfrac{a^m}{a^n} = a^{m-n}, \; m > n, \; a \neq 0$
3. $\dfrac{a^m}{a^n} = \dfrac{1}{a^{n-m}}, \; m < n, \; a \neq 0$
4. $(ab)^m = a^m b^m$
5. $\left(\dfrac{a}{b}\right)^m = \dfrac{a^m}{b^m}, \; b \neq 0$
6. $(a^m)^n = a^{mn}$

EXERCISES 2.1

Give the simplest number name involving no exponents (Exercises 1–12).

1. 2^3
2. 3^4
3. $(-1)^7$
4. $-(-5)^2$
5. $(-\tfrac{1}{2})^3$
6. $(10)^2$
7. $(10)^3$
8. $(10)^4$
9. $(0.1)^2$
10. 2^4
11. $-(2^4)$
12. $(-2)^4$

13. Tell whether each of the following names a positive or a negative number.
    (a) 2^4
    (b) 3^2
    (c) $(-1)^4$
    (d) $(-1)^3$
    (e) $(-17)^7$
    (f) $(125)^4$
    (g) $(-10)^{500}$
    (h) $(-18)^3(-18)^5$
    (i) $(-3)^7(-3)^6$
    (j) $(-25)^3(-25)^7$

Perform the indicated operations, using the laws of exponents. Assume that no divisors are zero (Exercises 14–53).

14. $x^7 \cdot x^3$
15. $a^5 \cdot a^2$
16. $r \cdot r^9$
17. $z^2 \cdot z^5$
18. $\dfrac{x^5}{x^3}$
19. $\dfrac{y^4}{y^7}$
20. $\dfrac{x^4 y^3}{xy^2}$
21. $(x^3)^3$
22. $(-x^5)^3$
23. $(x^2 y^2 z^3)^2$
24. $(2x^3 y^2)^3$
25. $\left(\dfrac{3x^3}{y}\right)^4$
26. $\left(\dfrac{-2x}{3y^2}\right)^3$
27. $\left(\dfrac{a^3}{bc^2}\right)^3$

EXPONENTS AND RADICALS

28. $\left(\dfrac{a^3}{b^2}\right)^2$

29. $\left(\dfrac{2y^2}{x^3}\right)^4$

30. $(m^2n^2)(-m^3n^2)^2$

31. $(4a^3)^3(2a^4)^3$

32. $\dfrac{(xy^2)^3}{(x^2y^3)^2}$

33. $\left(\dfrac{2a}{b}\right)^3\left(\dfrac{-3a^2}{b^3}\right)^4$

34. $\left(\dfrac{-4x^2y}{2xy}\right)^2$

35. $\left(\dfrac{2a^2}{x^3}\right)^3\left(\dfrac{x^2}{2a^3}\right)^4$

36. $\left(\dfrac{x}{y}\right)^2\left(\dfrac{-3x^2}{y^2}\right)^3\left(\dfrac{4y^2}{x^3}\right)^2$

37. $\left(\dfrac{-2}{y}\right)^3(2y^4)^2$

38. $\left(\dfrac{x}{y}\right)^3\left(\dfrac{-x}{y}\right)^4\left(\dfrac{y^7}{x^3}\right)^7$

39. $\left(\dfrac{-4x}{y}\right)^3\left(\dfrac{-3xy}{z}\right)^2\left(\dfrac{z^2}{y^2}\right)^8$

40. $x^n \cdot x^r$

41. $(x^r)^{4n}$

42. $\dfrac{x^{3n}}{x^{7n}}$

43. $\dfrac{y^{10k}}{y^k}$

44. $\dfrac{x^{2n}y^n}{x^{3n}y^n}$

45. $\dfrac{x^{a+1}}{x^{a-1}}$

46. $\dfrac{(3a^x)^3}{4a^{2x}}$

47. $\dfrac{x^{a+b}}{x^{a-b}}$

48. $\left(\dfrac{x^{3a}}{x^b}\right)^2\left(\dfrac{x^{5b}}{x^{4a}}\right)^3$

49. $\left(\dfrac{x^{n+1}}{x^n}\right)^n$

50. $\dfrac{(a-b)^4(a+b)^4}{(a-b)^3(a+b)^8}$

51. $\dfrac{(a+b)^4(a-b)^4}{(a-b)^8}$

52. $\dfrac{(a+b)^n(a-b)^{3k}}{(a+b)^{2n}(a-b)^k}$

53. $\dfrac{(y^{n+1}y^{2n+1})^2}{y^{4n}}$

2.2 ◆ Roots and Radicals

Let us consider the equation

$$x^2 = 25$$

This equation has two solutions: -5 and 5, since $(-5)^2 = 25$ and $5^2 = 25$. The real numbers -5 and 5 are then **square roots** of 25, since $(5)(5) = 25$ and $(-5)(-5) = 25$. That is, each of the real numbers -5 and 5 is a square root of 25. *Every positive real number has two square roots: one a negative real number and the other a positive real number.* In order to distinguish between these, we call the positive square root the **principal square root**.

Whenever x is positive, we use $\sqrt{x}$ to denote the principal square root of x. That is, we interpret the symbol $\sqrt{}$ to mean "the positive number whose square is." Since we also know that $0^2 = 0$, we extend the above

2.2 ROOTS AND RADICALS

definition of the symbol $\sqrt{}$ to say "the nonnegative number whose square is." If we wish to indicate the negative number that is a square root of x, we must write $-\sqrt{x}$, that is, the opposite of the positive square root (the principal square root). Since the square of a real number is never negative, the symbol $\sqrt{x}$ has meaning only if $x \geq 0$. We then have a basic rule governing square roots.

NEVER TAKE A SQUARE ROOT OF A NEGATIVE NUMBER.

We shall now state without proof some properties of the operation of taking the (principal) square root of a nonnegative number.

1. If $x \geq 0$ and $y \geq 0$, then $\sqrt{xy} = \sqrt{x} \cdot \sqrt{y}$.
2. If $x \geq 0$ and $y > 0$, then $\sqrt{\dfrac{x}{y}} = \dfrac{\sqrt{x}}{\sqrt{y}}$.
3. If a is any real number, then $\sqrt{a^2} = |a|$.

We have defined the square root of a nonnegative number as one of the two equal nonnegative factors of that number. We now extend this definition to cover the cases of the cube root, the fourth root, and, in general, the nth root, where n is any positive integer.

We define the **cube root** of a real number as one of the three equal factors of that number. Thus, since $8 = 2 \cdot 2 \cdot 2$, 2 is the cube root of 8. We write $\sqrt[3]{8} = 2$. Also, since $-8 = (-2)(-2)(-2)$, we have $\sqrt[3]{-8} = -2$. Since the cube root of a negative real number is negative, we are not restricted in taking cube roots as we are in taking square roots.

The **fourth root** of a real number is one of its four equal nonnegative factors, and, in general, the **nth root** of a real number is one of its n equal factors. It must be emphasized that we can only find an even root of a nonnegative number, and we take that even root to be nonnegative.

The nth root of the real number x is denoted by $\sqrt[n]{x}$ where n is called the **index** and x is called the **radicand.**

The principal even root of a nonnegative number is nonnegative as is an odd root of a nonnegative number. An odd root of a negative number is negative. It is not possible to find an even root of a negative number.

We now extend, without proof, the properties of square roots to the general case of the nth root.

1. $\sqrt[n]{xy} = \sqrt[n]{x} \cdot \sqrt[n]{y}$.
2. $\sqrt[n]{\dfrac{x}{y}} = \dfrac{\sqrt[n]{x}}{\sqrt[n]{y}}, y \neq 0$.
3. If n is even, $\sqrt[n]{a^n} = |a|$.
   If n is odd, $\sqrt[n]{a^n} = a$.

EXPONENTS AND RADICALS

EXERCISES 2.2

Simplify each of the following.

1. $\sqrt{64}$
2. $-\sqrt{25}$
3. $\sqrt[3]{-27}$
4. $-\sqrt[4]{16}$
5. $\sqrt[4]{x^4}$
6. $\sqrt[3]{-x^3y^9}$
7. $\sqrt[5]{x^5y^5}$
8. $\sqrt[3]{27y^6}$
9. $-\sqrt[5]{-32}$
10. $\sqrt{0.04}$
11. $\sqrt{\frac{1}{16}x^{16}}$
12. $\sqrt[3]{8x^6y^{12}}$
13. $\sqrt{x^6}$
14. $\sqrt[6]{64}$
15. $-\sqrt{144}$
16. $\sqrt[3]{\frac{8}{27}x^6y^9}$
17. $\sqrt[5]{32x^{10}}$
18. $-\sqrt[9]{x^9y^{18}}$
19. $\sqrt[4]{81x^4y^8}$
20. $\sqrt{\frac{25}{16}x^8y^6}$
21. $\sqrt[3]{\frac{-8}{125}y^6}$
22. $-\sqrt[3]{-27x^3}$
23. $\sqrt{\frac{144}{289}y^4}$
24. $\sqrt[3]{\frac{-125y^3}{x^6}}$
25. $\sqrt[5]{\frac{-32x^{10}}{y^{10}z^5}}$

2.3 ◆ Positive Rational Number Exponents

We have defined a^n where n is a positive integer. We now define a^r where r is a positive rational number—that is, a number of the form $\frac{p}{q}$ where p and q are positive integers. We shall define rational number powers so that they will be governed by the same laws of exponents as positive integer powers. For convenience, we restate the laws for positive integer exponents.

LAW 1. If m and n are positive integers and a is any real number, then

$$a^m a^n = a^{m+n}$$

LAW 2. If m and n are positive integers, $m > n$, and $a \neq 0$, then

$$\frac{a^m}{a^n} = a^{m-n}$$

LAW 3. If m and n are positive integers, $m < n$, and $a \neq 0$, then

$$\frac{a^m}{a^n} = \frac{1}{a^{n-m}}$$

LAW 4. If m and n are positive integers and a is any real number, then

$$(a^m)^n = a^{mn}$$

LAW 5. If n is a positive integer and a and b are any real numbers, then

$$(ab)^n = a^n \cdot b^n$$

2.3 POSITIVE RATIONAL NUMBER EXPONENTS

LAW 6. If n is a positive integer, a and b are real numbers, and $b \neq 0$, then

$$\left(\frac{a}{b}\right)^n = \frac{a^n}{b^n}$$

We want positive rational number powers to obey the same laws of exponents as positive integer powers. Law 4 shows that for $a \geq 0$ we want

$$(a^{1/2})^2 = a^{(1/2) \cdot 2} = a^1 = a$$

We see that if we define

$$a^{1/2} = \sqrt{a}$$

when $a \geq 0$, then Law 4 will be satisfied. If n is a positive integer, we define

$$a^{1/n} = \sqrt[n]{a}$$

This means that if n is even, we can only have the base, a, as a nonnegative number, while if n is odd, we can have any real number for the base.

Since we have defined $a^{1/n}$ as $\sqrt[n]{a}$, we see, for example, that

$$a^{2/3} = (a^{1/3})^2 = (\sqrt[3]{a})^2$$

and

$$a^{2/3} = (a^2)^{1/3} = \sqrt[3]{a^2}$$

In general, if m and n are positive integers, we define

$$a^{m/n} = (\sqrt[n]{a})^m = \sqrt[n]{a^m}$$

EXAMPLES

$27^{2/3} = (\sqrt[3]{27})^2 = 3^2 = 9$ and $27^{2/3} = \sqrt[3]{27^2} = \sqrt[3]{729} = 9$

$(x^2 y^4)^{1/2} = (x^2)^{1/2}(y^4)^{1/2} = \sqrt{x^2}\sqrt{y^4} = |x|y^2$

$\left(\dfrac{x^{2/3}}{y^{1/3}}\right)^6 = \dfrac{(x^{2/3})^6}{(y^{1/3})^6} = \dfrac{x^{(2/3)(6)}}{y^{(1/3)(6)}} = \dfrac{x^4}{y^2}$

$\left(\dfrac{x^6 y^3}{x^3 y^9}\right)^{1/3} = \left(\dfrac{x^3}{y^6}\right)^{1/3} = \dfrac{x}{y^2}$

EXERCISES 2.3

Find the value of each of the following (Exercises 1–12).

1. $8^{1/3}$
2. $(27)^{2/3}$
3. $4^{1/2}$
4. $(-8)^{2/3}$
5. $(0.04)^{1/2}$
6. $(-1)^{1/3}$

34   EXPONENTS AND RADICALS

7. $(\frac{8}{27})^{2/3}$ 9. $9^{3/2}$ 11. $(81)^{1/2}$
8. $(81)^{1/4}$ 10. $(16)^{3/4}$ 12. $(-\frac{125}{8})^{2/3}$

Write each of the following in radical form (Exercises 13–40).

13. $x^{1/2}$ 27. $(8x)^{2/3}$
14. $a^{2/3}$ 28. $(3b^3)^{4/3}$
15. $x^{3/8}$ 29. $(x^2y)^{1/4}$
16. $c^{3/4}$ 30. $2(9x)^{1/3}$
17. $5^{2/3}$ 31. $(x^3y^4)^{2/3}$
18. $(-3)^{4/3}$ 32. $(x^2y^5)^{3/8}$
19. $(ax)^{1/3}$ 33. $4x^2(x^3y^4)^{3/5}$
20. $(3x)^{1/2}$ 34. $(x-y)^{1/2}$
21. $(ab)^{3/4}$ 35. $(x+y)^{2/3}$
22. $(2x)^{5/2}$ 36. $(x^2-y^2)^{3/4}$
23. $(10x)^{3/2}$ 37. $(3x+y)^{2/3}$
24. $10 \cdot x^{3/2}$ 38. $(a-3b)^{1/4}$
25. $3 \cdot x^{1/2}$ 39. $(2x+y)^{3/5}$
26. $8 \cdot x^{2/3}$ 40. $(x^4-2y^4)^{1/5}$

Write each of the following with rational exponents (Exercises 41–60).

41. $\sqrt{8}$ 51. $\sqrt[3]{-x^2}$
42. $\sqrt{x}$ 52. $\sqrt[5]{-32}$
43. $\sqrt[3]{5}$ 53. $\sqrt[3]{-37x^4}$
44. $\sqrt[4]{25}$ 54. $\sqrt[3]{x^2y^2z^4}$
45. $\sqrt[3]{y^2}$ 55. $\sqrt[4]{z^3y^5}$
46. $\sqrt[5]{(2a)^3}$ 56. $\sqrt[6]{x^3y^3}$
47. $\sqrt[3]{3x^2}$ 57. $-2\sqrt{10x}$
48. $\sqrt[8]{y^7}$ 58. $\sqrt[5]{(x-y)^2}$
49. $\sqrt[3]{z^2}$ 59. $\sqrt[3]{x}\sqrt[3]{y}$
50. $\sqrt[3]{-8a^2}$ 60. $\sqrt{x}-\sqrt{y}$

Simplify (Exercises 61–82).

61. $(x^{1/2}y^{1/2})^2$ 69. $y^{2/3} \cdot y^{1/3}$
62. $(a^{1/4}b^{1/4})^3$ 70. $(a^{2/3}b^{1/3})^3$
63. $x^{1/2} \cdot x^{1/2}$ 71. $(-8x^3y^2)^{1/3}$
64. $a^{1/2}a^{3/4}$ 72. $\dfrac{x^{3/2}}{x^{1/2}}$
65. $2 \cdot 2^{1/2} \cdot 2^{1/4}$
66. $(9a^2b^4)^{1/2}$ 73. $\dfrac{y^{1/3}}{y^{2/3}}$
67. $(x^{1/2})^6$
68. $x^{1/4} \cdot x^{2/3}$

2.4 ZERO AND NEGATIVE RATIONAL EXPONENTS

74. $\left(\dfrac{c^{2/3}}{c^{1/3}}\right)^4$

75. $\left(\dfrac{a^7 b^3}{a^2 b^6}\right)^{1/3}$

76. $\left(\dfrac{x^{3/4}}{x^{1/2}}\right)^3$

77. $\left(\dfrac{x^6 y^3}{x^{12} y^7}\right)^{1/3}$

78. $\left(\dfrac{a^{2/3} b^{2/3}}{a^{1/3} a^{1/3}}\right)^3$

79. $\dfrac{(p^{3/4} r^{1/2})^4}{(-p^{1/3} r^{2/3})^3}$

80. $\dfrac{(-2 x^{1/6} y^{1/2})^5}{(-x^{1/4} y^{1/4})^4}$

81. $\left(\dfrac{-12 x^{1/3} y^{5/6}}{3 x^{2/3} y^{1/6}}\right)^4$

82. $\left(\dfrac{-3 x^{1/3} y^{3/4} z^{1/2}}{12 x^{2/3} y^{5/4} z}\right)^5$

2.4 • Zero and Negative Rational Exponents

We now define a **zero exponent** and **negative rational exponents** so that the laws of exponents that concern them will be consistent with those for the positive rational exponents.

From Law 2 we have, for $a \neq 0$ and m a positive rational number,

$$\dfrac{a^m}{a^m} = a^{m-m} = a^0$$

But, when $a \neq 0$, $\dfrac{a^m}{a^m} = 1$, so we are led to define the zero power of any real number to be 1. (Although we may not divide by zero, we may extend the definition to have $0^0 = 1$, since raising to the zero power does not necessarily involve division.) Therefore,

$$a^0 = 1$$

Thus

$$3^0 = 1$$
$$x^0 = 1$$
$$(x + y)^0 = 1$$

Using the definition $a^0 = 1$, we have, for n a positive rational number and $a \neq 0$,

$$\dfrac{1}{a^n} = \dfrac{a^0}{a^n} = a^{0-n} = a^{-n}$$

Thus, for n a positive rational number,

$$a^{-n} = \dfrac{1}{a^n}$$

EXPONENTS AND RADICALS

EXAMPLES

$$7^{-2} = \frac{1}{7^2} = \frac{1}{49}$$

$$3^{-1/3} = \frac{1}{3^{1/3}} = \frac{1}{\sqrt[3]{3}}$$

$$(a+b)^{-1} = \frac{1}{a+b}, \quad a \neq -b$$

We note that, if $a \neq 0$, $b \neq 0$, and n is a positive rational number, then

$$\frac{a^{-n}}{b^{-n}} = \frac{\frac{1}{a^n}}{\frac{1}{b^n}} = \frac{1}{a^n} \cdot \frac{b^n}{1} = \frac{b^n}{a^n}$$

From this we can conclude that any factor in the numerator of a fraction may be placed as a factor in the denominator if the sign of the exponent is changed. Likewise, any factor in the denominator of a fraction may be placed as a factor in the numerator if the sign of the exponent is changed. It must be remembered that *factors* and *not terms* may be changed by this rule. For example:

$$\frac{3x^3 y^{-2}}{xy^2 z^{-3}} = \frac{3x^3 z^3}{xy^4} = \frac{3x^2 z^3}{y^4}$$

$$x^2 y^{-3} = \frac{x^2}{y^3}$$

$$x + y^{-1} = x + \frac{1}{y}$$

$$\frac{x^{-1} + y^{-1}}{x^{-1} - y^{-1}} = \frac{\frac{1}{x} + \frac{1}{y}}{\frac{1}{x} - \frac{1}{y}} = \frac{y+x}{y-x}$$

EXERCISES 2.4

Write each of the following in a form in which all the exponents are positive (Exercises 1–20).

1. x^{-2}
2. y^{-3}
3. $z^{-1/3}$
4. $x^{-2} y^{-5}$
5. $a^{-4} b$
6. $x^3 y^{-4}$

2.5 CHANGING THE FORM OF RADICALS

7. $\left(\frac{2}{3}\right)^{-1}$
8. $\dfrac{1}{x^{-2}y^{-3}}$
9. $\dfrac{2}{c^{-3}d^8}$
10. $\dfrac{x^{-2}}{y^{-3}}$
11. $\dfrac{3x^2y^{-3}}{z^{-2}t^3}$
12. $\dfrac{x^3y^{-4}}{3x^4y^{-6}}$
13. $\dfrac{(8a)^{-1}}{(3b)^{-2}}$
14. $\dfrac{n^{-2}m^6}{3x^{-4}k^k}$
15. $x^{-1} + y^{-1}$
16. $3a^{-1} + b^{-2}$
17. $\dfrac{x + x^{-1}}{y + y^{-1}}$
18. $\dfrac{2a^{-1} - 2b^{-1}}{a^{-2} - b^{-2}}$
19. $\dfrac{a^{-1} - b^{-1}}{2a^{-1} - 2b^{-1}}$
20. $\dfrac{x^{-2} - y^{-2}}{x^{-2} + y^{-2}}$

Simplify. Write answers containing only positive exponents (Exercises 21–40).

21. $(4x^0)(4x)^0$
22. $(2^2)(2^{-3})$
23. $(3x^2y)(2x^{-1}y^2)$
24. $(-5x^2y^{-2})^3$
25. $(x^2y^3)(x^{-1}y^{-4})$
26. $(x^{1/2})^{-3}$
27. $(x^{1/2})(x^{-1/4})$
28. $2(x^{-2}y)(x^2y^{-4})$
29. $(3x^{1/2})(-2x^{5/2})$
30. $(a^{-2}b^{-4}c^{-3})(3abc)$
31. $(8x^2)(24x)^{-2}$
32. $\dfrac{x^{1/2}y^{2/3}z^{-3}}{x^{-1/2}y^{-1/3}z^4}$
33. $\left(\dfrac{2x^{-1}y}{6x^3y^{-4}}\right)^2$
34. $\dfrac{(16d^3)^{-1}}{(4d^5)^{-2}}$
35. $\left(\dfrac{8x^{-6}}{27b^3}\right)^{-2/3}$
36. $\dfrac{(-3x^{1/3}y^{1/4})^{12}}{(12x^{1/4}y^{1/2})^8}$
37. $\left(\dfrac{x^n}{x^{n-1}}\right)^{-1}$
38. $\left(\dfrac{x^n}{x^{n-1}}\right)^{-3}$
39. $\left(\dfrac{x^{3n}y^{4n}}{x^{2n}y^n}\right)^{-3}$
40. $\left(\dfrac{x^{2n}y^{n-1}}{x^{n-1}y^2}\right)^{-3}$

2.5 ◆ Changing the Form of Radicals

We know that

$$\sqrt[n]{ab} = \sqrt[n]{a}\sqrt[n]{b} \quad \text{and} \quad \sqrt[n]{\dfrac{a}{b}} = \dfrac{\sqrt[n]{a}}{\sqrt[n]{b}}$$

whenever these roots exist and we do not divide by 0. We now use these facts to *simplify* expressions containing radicals. That is, we put them in a form that is more convenient for computation.

38   EXPONENTS AND RADICALS

There are three methods used to simplify radical expressions. They are as follows.

1. Remove from the radicand any factor that is raised to a power equal to the index or to an integral multiple of the index.

(a) $\sqrt{8} = \sqrt{4 \cdot 2} = \sqrt{4} \cdot \sqrt{2} = \sqrt{2^2} \cdot \sqrt{2} = 2\sqrt{2}$

(b) $\sqrt[3]{16a^5} = \sqrt[3]{2^4 a^5} = \sqrt[3]{(2^3 a^3)(2a^2)} = \sqrt[3]{2^3 a^3} \sqrt[3]{2a^2} = 2a\sqrt[3]{2a^2}$

(c) $2\sqrt{32x^2 y^4} = 2\sqrt{2^5 x^2 y^4} = 2\sqrt{(2^4 x^2 y^4)(2)} = 2\sqrt{2^4 x^2 y^4} \cdot \sqrt{2}$
$= 2(2^2 |x| y^2)\sqrt{2} = 8|x| y^2 \sqrt{2}$

(d) $\sqrt[3]{-16x^3 y^7} = \sqrt[3]{(-8x^3 y^6)(2y)} = \sqrt[3]{(-8x^3 y^6)} \cdot \sqrt[3]{2y} = -2xy^2 \sqrt[3]{2y}$

2. Remove from the radicand any fractional factor. To do this we rename the fraction as an equivalent fraction whose denominator is the square root of a perfect square, the cube root of a perfect cube, or the nth root of an nth power. This is called **rationalizing the denominator,** since the result is a denominator that contains only rational real numbers or rational algebraic expressions.

(a) $\sqrt{\dfrac{1}{2}} = \sqrt{\dfrac{1 \cdot 2}{2 \cdot 2}} = \sqrt{\dfrac{2}{4}} = \dfrac{\sqrt{2}}{\sqrt{4}} = \dfrac{\sqrt{2}}{2} = \dfrac{1}{2}\sqrt{2}$

(b) $\sqrt[3]{\dfrac{4}{3}} = \sqrt[3]{\dfrac{4 \cdot 3^2}{3 \cdot 3^2}} = \dfrac{\sqrt[3]{4 \cdot 9}}{\sqrt[3]{3^3}} = \dfrac{\sqrt[3]{36}}{3} = \dfrac{1}{3}\sqrt[3]{36}$

(c) $\sqrt[4]{\dfrac{3a^2}{8b^2}} = \sqrt[4]{\dfrac{3a^2 \cdot 2b^2}{2^3 b^2 \cdot 2b^2}} = \dfrac{\sqrt[4]{6a^2 b^2}}{\sqrt[4]{2^4 b^4}} = \dfrac{\sqrt[4]{6a^2 b^2}}{2|b|}$

(d) $\dfrac{a}{\sqrt{a} + \sqrt{b}} = \dfrac{a(\sqrt{a} - \sqrt{b})}{(\sqrt{a} + \sqrt{b})(\sqrt{a} - \sqrt{b})} = \dfrac{a(\sqrt{a} - \sqrt{b})}{(\sqrt{a})^2 - (\sqrt{b})^2}$
$= \dfrac{a(\sqrt{a} - \sqrt{b})}{a - b}$

3. Reduce the order of the index.

(a) $\sqrt[4]{4} = \sqrt[4]{2^2} = 2^{2/4} = 2^{1/2} = \sqrt{2}$

(b) $\sqrt[6]{x^2} = x^{2/6} = x^{1/3} = \sqrt[3]{x}$

(c) $\sqrt[6]{27a^3} = \sqrt[6]{3^3 a^3} = 3^{3/6} \cdot a^{3/6} = 3^{1/2} \cdot a^{1/2} = \sqrt{3a}$

We say that radicals obtained by the application of the three methods above are in **simplest form.** An expression involving radicals will be in simplest form when:

1. The radicand contains no fractions.

2.5 CHANGING THE FORM OF RADICALS

2. The radicand contains no factors that are perfect nth powers, where n is the index of the radical.
3. The index is the smallest possible positive integer.

EXAMPLE 1. Simplify $\sqrt{54}$.

SOLUTION
$$\sqrt{54} = \sqrt{9 \cdot 6} = \sqrt{9}\sqrt{6} = 3\sqrt{6}$$

EXAMPLE 2. Simplify $\sqrt[3]{\dfrac{y}{2x}}$, $x \neq 0$.

SOLUTION. We multiply numerator and denominator of the radicand by $2^2 x^2$ to obtain an equivalent fraction whose denominator is a perfect cube:

$$\sqrt[3]{\frac{y}{2x}} = \sqrt[3]{\frac{y \cdot 2^2 \cdot x^2}{(2x)(2^2 x^2)}} = \frac{\sqrt[3]{4x^2 y}}{\sqrt[3]{2^3 x^3}} = \frac{\sqrt[3]{4x^2 y}}{2x} = \frac{1}{2x}\sqrt[3]{4x^2 y}$$

EXAMPLE 3. Simplify $\sqrt[6]{216}$.

SOLUTION
$$\sqrt[6]{216} = \sqrt[6]{6^3} = 6^{3/6} = 6^{1/2} = \sqrt{6}$$

EXAMPLE 4. Simplify $2\sqrt[3]{56x^2 y^3}$.

SOLUTION
$$2\sqrt[3]{56x^2 y^3} = 2\sqrt[3]{(8y^3)(7x^2)}$$
$$= 2\sqrt[3]{8y^3}\sqrt[3]{7x^2}$$
$$= 2(2y)\sqrt[3]{7x^2}$$
$$= 4y\sqrt[3]{7x^2}$$

EXAMPLE 5. Rationalize the denominator of $\dfrac{7}{\sqrt{5} - \sqrt{2}}$.

SOLUTION. We use the fact that $(a + b)(a - b) = a^2 - b^2$:

$$(\sqrt{5} - \sqrt{2})(\sqrt{5} + \sqrt{2}) = (\sqrt{5})^2 - (\sqrt{2})^2 = 5 - 2 = 3$$

Then
$$\frac{7}{\sqrt{5}-\sqrt{2}} = \frac{7(\sqrt{5}+\sqrt{2})}{(\sqrt{5}-\sqrt{2})(\sqrt{5}+\sqrt{2})}$$
$$= \frac{7\sqrt{5}+7\sqrt{2}}{(\sqrt{5})^2-(\sqrt{2})^2}$$
$$= \frac{7\sqrt{5}+7\sqrt{2}}{5-2} = \frac{7\sqrt{5}+7\sqrt{2}}{3}$$

EXERCISES 2.5

Simplify (Exercises 1–16).

1. $\sqrt{16}$
2. $\sqrt{12}$
3. $\sqrt{4y}$
4. $\sqrt{20}$
5. $\sqrt{xy^2}$
6. $\sqrt{32}$
7. $\sqrt{40x^2}$
8. $\sqrt{54a^2}$
9. $\sqrt[3]{y^3}$
10. $\sqrt[3]{64}$
11. $\sqrt{20x^4}$
12. $\sqrt{45x^3y^6}$
13. $\sqrt[3]{-27xy^4}$
14. $\sqrt{32x^5y^7}$
15. $\sqrt{100xy^3}$
16. $\sqrt[3]{-81x^4y^3}$

Rationalize the denominators (Exercises 17–26).

17. $\sqrt{\frac{1}{3}}$
18. $\sqrt{\frac{2}{3}}$
19. $\frac{x}{\sqrt{x}}$
20. $-\sqrt{\frac{1}{2x}}$
21. $\frac{2}{\sqrt[3]{4}}$
22. $\sqrt{\frac{1}{n}}$
23. $\sqrt{\frac{2}{x}}$
24. $\frac{1}{\sqrt[3]{a^2b^2}}$
25. $\sqrt{\frac{ab}{y}}$
26. $\frac{\sqrt{x^2+y^2}}{\sqrt{x}}$

Reduce the order of the index of each radical (Exercises 27–36).

27. $\sqrt[6]{8}$
28. $\sqrt[4]{9}$
29. $\sqrt[4]{y^2}$
30. $\sqrt[6]{x^3}$
31. $\sqrt[6]{27a^3}$
32. $\sqrt[4]{4a^2}$
33. $\sqrt[4]{49a^2}$
34. $\sqrt[6]{27x^3y^3}$
35. $\sqrt[10]{y^5z^{10}}$
36. $\sqrt[4]{16a^2b^4}$

2.6 OPERATIONS CONTAINING RADICAL EXPRESSIONS

Simplify each of the following (Exercises 37–46).

37. $\sqrt[3]{108x^5}$
38. $2\sqrt{48x^3}$
39. $\sqrt[6]{49x^4}$
40. $\sqrt[3]{81x^4y^6z^5}$
41. $\sqrt[4]{\dfrac{64x^2y^4}{z^2}}$
42. $\sqrt{192x^3y^7}$
43. $\sqrt[4]{\dfrac{x^7y^2}{27}}$
44. $\sqrt{x^{-2} + y^{-2}}$
45. $\sqrt[3]{\dfrac{125x^4y^3}{4}}$
46. $\sqrt{x^{-1} + y^{-1}}$

Rationalize the denominators (Exercises 47–50).

47. $\dfrac{2}{\sqrt{5} + \sqrt{3}}$
48. $\dfrac{6}{\sqrt{2} + \sqrt{6}}$
49. $\dfrac{8}{5\sqrt{3} - 7}$
50. $\dfrac{12}{4 + 3\sqrt{6}}$

2.6 ◆ Operations Containing Radical Expressions

Radicals are said to be **similar** if they have the same radicand and the same index. Thus $\sqrt{2}$ and $\sqrt{18} = 3\sqrt{2}$ are similar. Since radical expressions are symbols for real numbers, the distributive property applies to them. We use the distributive property when we add and subtract similar radicals. Thus

$$2\sqrt{3} + 4\sqrt{3} = (2 + 4)\sqrt{3} = 6\sqrt{3}$$
$$5\sqrt{5} - 8\sqrt{5} = (5 - 8)\sqrt{5} = -3\sqrt{5}$$
$$\sqrt{8} + \sqrt{18} = 2\sqrt{2} + 3\sqrt{2} = (2 + 3)\sqrt{2} = 5\sqrt{2}$$
$$3\sqrt{48} - 4\sqrt{12} = 12\sqrt{3} - 8\sqrt{3} = (12 - 8)\sqrt{3} = 4\sqrt{3}$$

Addition and subtraction of radicals that are not similar can only be indicated. Thus the simplest form of the sum of $\sqrt{2}$ and $\sqrt{3}$ is $\sqrt{2} + \sqrt{3}$.

The laws of exponents for positive rational exponents assure us that

$$x^{1/n} \cdot y^{1/n} = (xy)^{1/n}$$

hence

$$\sqrt[n]{x}\sqrt[n]{y} = \sqrt[n]{xy}$$

From the above we see that two radicals of the same **order**, that is, with the same index may be expressed as a single radical whose radicand is the product of the radicands involved and whose index is the same as the index of the factors. Thus

$$\sqrt{2} \cdot \sqrt{3} = \sqrt{2 \cdot 3} = \sqrt{6}$$
$$(2\sqrt{3})(3\sqrt{5}) = (2)(3)\sqrt{3 \cdot 5} = 6\sqrt{15}$$
$$(3\sqrt{5})(4\sqrt{6}) = (3)(4)\sqrt{5 \cdot 6} = 12\sqrt{30}$$

We use the distributive property when finding the product of expressions containing radicals. Note the use of the distributive property in the examples below.

$$x(\sqrt{3} + \sqrt{5}) = x\sqrt{3} + x\sqrt{5}$$
$$2\sqrt{2}(\sqrt{3} + 4\sqrt{5}) = 2\sqrt{6} + 8\sqrt{10}$$
$$(\sqrt{2} + \sqrt{3})(\sqrt{2} - \sqrt{3}) = (\sqrt{2} + \sqrt{3})\sqrt{2} + (\sqrt{2} + \sqrt{3})(-\sqrt{3})$$
$$= 2 + \sqrt{6} - \sqrt{6} - 3$$
$$= 2 - 3$$
$$= -1$$

$$(3\sqrt{5} + 2\sqrt{2})(2\sqrt{5} + 3\sqrt{2}) = (3\sqrt{5} + 2\sqrt{2})(2\sqrt{5})$$
$$+ (3\sqrt{5} + 2\sqrt{2})(3\sqrt{2})$$
$$= 3 \cdot 2 \cdot 5 + 4\sqrt{10} + 9\sqrt{10} + 2 \cdot 3 \cdot 2$$
$$= 30 + 4\sqrt{10} + 9\sqrt{10} + 12$$
$$= 42 + 13\sqrt{10}.$$

The quotient of two radical expressions having the same index is a radical with the same index and whose radicand is equal to the quotients of the two radicands involved. Thus, if $b \neq 0$ and n is a positive integer,

$$\frac{\sqrt[n]{a}}{\sqrt[n]{b}} = \sqrt[n]{\frac{a}{b}}$$

EXAMPLES

$$\frac{\sqrt{6}}{\sqrt{2}} = \sqrt{\frac{6}{2}} = \sqrt{3}$$

$$\frac{12\sqrt[3]{4b^2}}{6\sqrt[3]{2b^3}} = \frac{12}{6}\sqrt[3]{\frac{4b^2}{2b^3}} = 2\sqrt[3]{\frac{2}{b}} = 2\sqrt[3]{\frac{2b^2}{b^3}} = \frac{2}{b}\sqrt[3]{2b^2}$$

2.6 OPERATIONS CONTAINING RADICAL EXPRESSIONS

EXERCISES 2.6

Perform the following operations. Write each answer in simplest form (Exercises 1–40).

1. $3\sqrt{2} + \sqrt{50} - 2\sqrt{18}$
2. $\sqrt{54} + \sqrt{150} - \sqrt{96}$
3. $2\sqrt{48} - 3\sqrt{27} + 5\sqrt{12}$
4. $\sqrt{112} + 5\sqrt{63} - \sqrt{175}$
5. $6\sqrt[3]{2} + \sqrt[3]{16} - \sqrt[3]{250}$
6. $2\sqrt[3]{40} + 3\sqrt[3]{135} - 4\sqrt[3]{320}$
7. $a\sqrt{27a} + \sqrt{108a^3} + 2\sqrt{3a^3}$
8. $3a\sqrt{b^5} - b\sqrt{4a^2b^3} - b^2\sqrt{16a^2b}$, $a \geq 0$
9. $x\sqrt{2x^3} + 8\sqrt{8x^5} - x^2\sqrt{98x}$
10. $\sqrt{3y^9} + y^2\sqrt{27y^5} + 2y^4\sqrt{108y}$
11. $\sqrt{\frac{1}{2}} + \sqrt{\frac{1}{8}} + \sqrt{\frac{9}{32}}$
12. $\sqrt{\frac{1}{12}} + \sqrt{\frac{1}{3}} - 5\sqrt{\frac{1}{27}}$
13. $\sqrt{\frac{4}{5}} - \sqrt{\frac{16}{25}} - \sqrt{\frac{9}{20}}$
14. $xy^2\sqrt[3]{4x^2} + xy\sqrt[3]{108x^2y^3} + \sqrt[3]{-32x^5y^6}$
15. $3\sqrt{2} \cdot 4\sqrt{8}$
16. $5\sqrt{5} \cdot \sqrt{80}$
17. $3\sqrt{3} \cdot \sqrt{12}$
18. $\sqrt{75} \cdot 2\sqrt{3}$
19. $\sqrt[3]{3} \cdot \sqrt[3]{9}$
20. $\sqrt[3]{24} \cdot \sqrt[3]{9}$
21. $\sqrt[4]{2} \cdot \sqrt[4]{8}$
22. $\sqrt{10} \cdot \sqrt{20}$
23. $\sqrt{3xyz} \cdot \sqrt{6x^3y^2} \cdot \sqrt{2x^5y^6}$
24. $\sqrt{7ab} \cdot \sqrt{14a^2b^3}$
25. $\sqrt{5ab^7} \cdot \sqrt{3abc} \cdot \sqrt{15a^3b}$
26. $\sqrt[4]{4a^2b^4} \cdot \sqrt[4]{8a^3b^5}$
27. $\sqrt[4]{8a^6b^7} \cdot \sqrt[4]{4b^4c^3}$
28. $\sqrt{15}(\sqrt{5} - \sqrt{3} - 2\sqrt{15})$
29. $\sqrt{2}(\sqrt{3} - \sqrt{6} + \sqrt{8})$
30. $(\sqrt{2} - \sqrt{3})(\sqrt{2} + \sqrt{3})$
31. $(\sqrt{6} + 3\sqrt{2})^2$
32. $(2\sqrt{5} + 3\sqrt{2})^2$
33. $(\sqrt{5} - \sqrt{3})^2$
34. $(4\sqrt{3} - 2\sqrt{2})^2$
35. $(\sqrt{2} + \sqrt{3})(3\sqrt{2} - 2\sqrt{3})$
36. $(4\sqrt{5} - 2\sqrt{3})(3\sqrt{5} - 4\sqrt{3})$
37. $(3\sqrt{5} - 4)^2$
38. $(5\sqrt{3} + \sqrt{6})(2\sqrt{3} - 3\sqrt{6})$
39. $(\sqrt{a+b} - \sqrt{a-b})^2$
40. $(2\sqrt{x+1} - 3\sqrt{4-x})^2$

Perform the indicated divisions. Write the answers in simplest form (Exercises 41–50).

41. $\dfrac{\sqrt{6}}{\sqrt{3}}$
42. $\dfrac{\sqrt{a^3b}}{\sqrt{ab^2}}$
43. $\dfrac{\sqrt{x^4y^5}}{\sqrt{2x^7y^3}}$
44. $\dfrac{\sqrt[3]{3a^3b^5}}{\sqrt[3]{-18a^5b^7}}$
45. $\sqrt[5]{\dfrac{64y^6z^4}{8y^7z}}$
46. $\dfrac{\sqrt[3]{xy^4}}{\sqrt[3]{6x^3y^7}}$
47. $\sqrt[5]{\dfrac{-32a^{10}}{ab^8}}$
48. $\sqrt[3]{\dfrac{8x^3y^3}{10xy^2}}$
49. $\sqrt{\dfrac{8c^6d^6}{28c^9d^4}}$
50. $\sqrt{\dfrac{a^2 - b^2}{a^2 - 2ab + b^2}}$

2.7 ♦ Scientific Notation

In the scientific application of mathematics, very small and very large numbers are used in calculations. For example:

> The speed of light is 29,978,300,000 centimeters per second.
> The wave length of a gamma ray is 0.00000000002 centimeter.
> A cubic centimeter of water contains 34,000,000,000,000,000,000,000 molecules.
> The wave length of visible blue light is 0.000045 centimeter.

Very large and very small numbers are most conveniently written with the aid of exponents in a form called **scientific notation**. For example:

$$32{,}000 = (3.2)(10)^4$$
$$31.86 = (3.186)(10)^1$$
$$0.016 = (1.6)(10)^{-2}$$
$$0.00002 = (2)(10)^{-5}$$

When a number is represented as the product of a number between 1 and 10 and an integral power of 10, we say that it is given in **scientific notation**. A number written in this way may be converted to standard notation by moving the decimal point in the first factor the number of places indicated by the exponent of 10: to the right if the exponent is positive and to the left if the exponent is negative.

Scientific notation simplifies numerical calculation. For example,

$$\frac{16.4(10)^3}{4(10)^4} = \frac{16.4}{4} \cdot \frac{(10)^3}{(10)^4} = 4.1(10)^{-1} = 0.41$$

EXERCISES 2.7

Express the following in scientific notation (Exercises 1–6).

1. 268,000
2. 126.8
3. 0.00000026
4. 27,860,000
5. 14,000,000
6. 5,860,000,000,000

Express the following in standard form (Exercises 7–12).

7. $3.2(10)^4$
8. $1.6(10)^8$
9. $3.16(10)^{-4}$
10. $6.7(10)^{-5}$
11. $9.36(10)^{-20}$
12. $1.1(10)^{10}$

CHAPTER 2 REVIEW

Express the numerals in the following in scientific notation (Exercises 13–16).

13. The area of the earth's surface is 197,000,000 square miles.
14. A light year is the distance light travels in one year. It is about 5,900,000,000,000 miles.
15. The planet Pluto is 4,300,000,000 miles from the sun.
16. Radar waves have photon energy of 0.0000000000000000002 erg.

Compute (Exercises 17–24).

17. $\dfrac{(10)^3(10)^5}{10^4}$

18. $\dfrac{(10)^3(10)^{-5}}{(10)^2(10)^{-4}}$

19. $\dfrac{(10)^{-3}(10)^{-5}}{(10)^{-8}(10)^3}$

20. $\dfrac{(10)^2(10)^{-3}(10)^5}{(10)^{-7}(10)^{-4}}$

21. $[2.4(10)^4][8.1(10)^6]$

22. $\dfrac{[4(10)^5][15(10)^2]}{8(10)^3}$

23. $\dfrac{5.87(10)^8}{3.60(10)^2}$

24. $\dfrac{(0.0024)(0.0038)}{(0.0036)(0.0019)}$

CHAPTER 2 REVIEW

Perform the indicated operations, using the laws of exponents, and simplify.

1. $\dfrac{(x^2y^3z)^4}{(xy^2z^3)^2}$

2. $\sqrt{\tfrac{36}{25}x^4y^2}$

3. $\sqrt[3]{\dfrac{-8x^6y^3}{z^9}}$

4. $(x^4y^2z^3)^{2/3}$

5. $\dfrac{x^{5/2}y^{4/3}z^{1/4}}{x^2yz}$

6. $\dfrac{(-3x^{1/3}y^{1/2})^3}{3x^{1/2}y^{1/2}}$

7. $(x^{-1/2}y^3z^{1/4})^{-2}$

8. $\dfrac{a^{-1}+b}{b^{-1}+a}$

9. $\dfrac{(a^{-1}-b^{-1})(c^{-1}+d^{-1})}{(a^{-1}+b^{-1})(c^{-1}-d^{-1})}$

10. $(x^{-2}y^{1/2}z^{-3})^{-2}$

11. $\sqrt[3]{\dfrac{2x^4}{z^5y^2}}$

12. $\sqrt[4]{16a^2b^{-3}}$

13. $(3\sqrt[3]{x^2yz})(2\sqrt[3]{xy^2z})$

14. $(-2x^3y^{1/3})^{-2}$

15. $(\sqrt{a}-3\sqrt{b})^2$

16. $(x^{1/3}+y^{1/3})(x^{2/3}-x^{1/3}y^{1/3}+y^{2/3})$

17. $(x^{1/4}+y^{1/4})(x^{1/4}-y^{1/4})$

18. $(x^3-x^{-3})^2$

19. $\dfrac{[2.3(10)^5][3.1(10)^2]}{1.7(10)^3}$

20. $[1.8(10)^{-3}][2(10)^4][3(10)^{-2}]^2$

CHAPTER 3

Polynomials

3.1 ◆ Definitions

The sum of the real numbers a and b is expressed as $a + b$. Their product is expressed as ab, $(a)(b)$, or $a \cdot b$. Any collection of numerals, variables, and signs of operation which form the name of a real number is called an **algebraic expression**. For example,

$$2x + 3y - c$$
$$a + b + c$$
$$a - b - 4c$$

are all algebraic expressions.

In an expression of the form $A + B + C + \cdots$, the symbols A, B, C, ... are called the **terms** of the expression. The terms of the expression $2x + 3y - c$ are $2x$, $3y$, and $-c$. If a term in an expression is the product of several factors, any one factor is called the **coefficient** of the product of the remaining factors. Thus, in the term $4xyz$, 4 is the coefficient of xyz, $4x$ is the coefficient of yz, $4xy$ is the coefficient of z, and so forth. In this book, we shall use the word coefficient to refer to the numerical factor of the term. Thus, in the term $4xyz$, we call 4 the coefficient. In the term xyz, we say that 1 is the coefficient, since $xyz = (1)(xyz)$.

Each of the following is an example of a **polynomial** in x.

(a) $2x^3 - 4x^2 + 3x + 2$
(b) $5x^2 - x + 7$
(c) $x + a$
(d) -2

If n is a positive integer, then any expression of the general form

$$a_0x^n + a_1x^{n-1} + a_2x^{n-2} + \cdots + a_{n-2}x^2 + a_{n-1}x + a_n$$

in which $a_0 \neq 0$ is called a **polynomial in x of degree n**. The polynomial (a) above is a polynomial in x of degree 3; polynomial (b) has degree 2; polynomial (c) has degree 1; and polynomial (d) has degree 0. The coefficients $a_0, a_1, a_2, \ldots, a_n$ represent real numbers. If a polynomial is of degree n, then the coefficient, a_0, of the power x^n is not zero.

A polynomial that consists of only one term is called a **monomial**; if it consists of two terms it is called a **binomial**; and if it consists of three terms it is called a **trinomial**. For example,

$$3, \quad x^3, \quad \text{and} \quad 5xy^2$$

are monomials.

$$x + 2, \quad x^2 - 1, \quad \text{and} \quad 4x^2 - y^2$$

are binomials.

$$x^2 - 2x + 1, \quad x^2y + y^2x + xy, \quad \text{and} \quad x^2 - 4xy + y^2$$

are trinomials.

We shall use symbols such as P, Q, and R, to denote polynomials in x. Thus $P = x^2 + 3x - 4$ and $Q = 8x^3 + 7x^2 - 2x + 9$.

3.2 ◆ Classification of Polynomials

Polynomials may be classified in various ways. In the previous section we classified polynomials according to degree and according to the number of terms. Thus:

$x^3 + 4x^2 + 3x$ is a third degree trinomial.
$x^2 - 1$ is a second degree binomial.
$5x^5$ is a fifth degree monomial.

Another method of classifying polynomials is according to the number of variables. The polynomials above are all polynomials in one variable. On the other hand:

$3xy$ is a polynomial in two variables.
$3x + 4y - 5z$ is a polynomial in three variables.

Still another method of classifying polynomials has to do with the numerical coefficients of the terms. If the numerical coefficients of the terms of a given polynomial are all integers, we say that the polynomial is a **polynomial over the integers**. If the numerical coefficients of the terms of a polynomial are elements of some specified set of numbers, then we say that the polynomial is a polynomial over that set of numbers. For example:

$3x^2 + 4x + 5$ is a polynomial over the integers.
$\frac{1}{2}x^3 + \frac{1}{4}x + 6$ is a polynomial over the rational numbers.
$\sqrt{2}x + 3.7$ is a polynomial over the real numbers.

Since the numerical coefficients of each of the polynomials above are real numbers, all of the polynomials given are polynomials over the real numbers. This is true since the set of integers and the set of rational numbers are subsets of the set of real numbers. Thus every polynomial over the integers is also a polynomial over the set of rational numbers and a polynomial over the set of real numbers. Every polynomial over the rational numbers is also a polynomial over the real numbers.

We shall consider polynomials in which the replacement set for the variable or variables is the set of real numbers. When we replace each variable in a polynomial by some real number, we determine a real number. This process is called **evaluating** the polynomial, or finding the **value** of the polynomial. For example, the polynomial

$$x^2 - 2$$

has value

$2^2 - 2 = 2$ when x is replaced by 2
$3^2 - 2 = 7$ when x is replaced by 3
$0^2 - 2 = -2$ when x is replaced by 0
$(-1)^2 - 2 = -1$ when x is replaced by -1

EXERCISES 3.1

1. Classify each of the following as monomial, binomial, or trinomial, and give the degree of each.
   (a) $x^3 - 3x + 1$ (d) $x^4 - 3x + 2$
   (b) $x^2 - 2x$ (e) $2x + 3$
   (c) 3 (f) $2x^2 - 4x + 0$

2. Classify each of the following as polynomials over the integers, over the rational numbers, or over the real numbers. Give the most descriptive classification in each case.
   (a) $x^2 + 3x - 2$ (d) $3\pi + x^2$
   (b) $x^2 - \frac{1}{2}x + \frac{1}{4}$ (e) $3x^2 - 4x + 7$
   (c) $x^3 - \sqrt{2}x^2 + 3$ (f) $\frac{1}{4}x^2 - \frac{1}{2}x + 7$

3.3 ADDITION AND SUBTRACTION OF POLYNOMIALS

3. Classify each of the following according to the number of variables.
   (a) $3xy + 2$
   (b) $x^2 + 4xy + y^2$
   (c) $y + 3x$
   (d) $x + 3y + 4z$
   (e) $2x^2 + 3xy + 4$
   (f) $3zxy + 3yx$

Evaluate the given polynomials for the specified replacements (Exercises 4–10).

4. $x + 2$, x replaced by 5
5. $y^2 - 3$, y replaced by 2
6. $3x^3 - 3x^2$, x replaced by -1
7. $\frac{1}{2}x - 2$, x replaced by -4
8. $x^3 - 9$, x replaced by -2
9. $2(x + 2)$, x replaced by 0
10. $\frac{1}{4}x + 3$, x replaced by 5

11. Determine the value of the polynomial $3x^2 - 2x + 4$ if x is replaced by:
    (a) 0
    (b) 2
    (c) -1
    (d) $\frac{1}{2}$
    (e) $-\frac{1}{4}$
    (f) -3

12. Determine the value of the polynomial $x^2 + \sqrt{2}x + 3$ if x is replaced by:
    (a) $\sqrt{2}$
    (b) $2\sqrt{2}$
    (c) 0
    (d) $-2\sqrt{2}$

Which of the following statements are true (Exercises 13–20)?

13. Every binomial is a polynomial.
14. Every polynomial over the real numbers is a polynomial over the integers.
15. $x^2 + 3x + 2$ is a polynomial of the second degree.
16. 7 is not a polynomial.
17. A trinomial is a polynomial with two terms.
18. Every polynomial is a trinomial.
19. $\dfrac{2}{x^2}$ is a polynomial.
20. Every polynomial is a binomial of degree three.

3.3 • Addition and Subtraction of Polynomials

When we add two polynomials we use the commutative and associative properties of addition and the distributive property. Consider the following examples.

EXAMPLE 1

$$2x + 5x = (2 + 5)x \quad \text{distributive property}$$
$$= 7x$$

EXAMPLE 2

$$2x^3 + (x^2 + 3x^3) = 2x^3 + (3x^3 + x^2) \quad \text{commutative property}$$
$$= (2x^3 + 3x^3) + x^2 \quad \text{associative property}$$
$$= (2 + 3)x^3 + x^2 \quad \text{distributive property}$$
$$= 5x^3 + x^2$$

EXAMPLE 3

$$(12x^2 + 5x) + (5x^2 + x) = (12x^2 + 5x) + (x + 5x^2)$$
$$= 12x^2 + [5x + (x + 5x^2)]$$
$$= 12x^2 + [(5x + x) + 5x^2]$$
$$= 12x^2 + [5x^2 + (5x + x)]$$
$$= (12x^2 + 5x^2) + (5x + x)$$
$$= (12 + 5)x^2 + (5 + 1)x$$
$$= 17x^2 + 6x$$

The reader should give reasons for each step in Example 3 above.

Usually we shorten the writing involved in the addition of polynomials by not showing each step. Thus

$$(2x^2 + 3x) + (5x^2 + 7x) = (2x^2 + 5x^2) + (3x + 7x) = 7x^2 + 10x$$

We often show the addition of polynomials in a vertical form similar to that used in the addition of positive integers:

$$2x^2 + 3x$$
$$5x^2 + 7x$$
$$\overline{7x^2 + 10x}$$

Since a polynomial over the real numbers has a real number value, when we replace the variables with real numbers, we define subtraction of polynomials in the same way we defined subtraction of real numbers. Thus, if P and Q are polynomials, then

$$P - Q = P + (-Q)$$

where $-Q$ is the **additive inverse** (that is, the **opposite**) of Q. Since Q and $-Q$ are additive inverses, we must have

$$Q + (-Q) = 0$$

For example, if $Q = x^2 + x - 1$, then

$$-Q = -(x^2 + x - 1) = -x^2 + (-x) + [-(-1)]$$
$$= -x^2 - x + 1$$

In general, if $Q = A + B + C + \cdots$, then $-Q = -A - B - C - \cdots$.
Consider the examples below.

EXAMPLE 4

$$(3x^2 + 2x) - (2x^2 + 5x) = (3x^2 + 2x) + [-(2x^2 + 5x)]$$
$$= (3x^2 + 2x) + [(-2x^2) + (-5x)]$$
$$= [3x^2 + (-2x^2)] + [2x + (-5x)]$$
$$= x^2 - 3x$$

3.3 ADDITION AND SUBTRACTION OF POLYNOMIALS

EXAMPLE 5

$$\begin{aligned}(2x^4 - 3x^2) - (x^4 - 5x^2) &= (2x^4 - 3x^2) + [-(x^4 - 5x^2)] \\ &= (2x^4 - 3x^2) + [(-x^4) + 5x^2] \\ &= x^4 + 2x^2\end{aligned}$$

EXAMPLE 6

$(4x^2 - 2x + 3) - (-x^2 - 2x + 2)$
$$\begin{aligned}&= (4x^2 - 2x + 3) + [-(-x^2 - 2x + 2)] \\ &= (4x^2 - 2x + 3) + [x^2 + 2x + (-2)] \\ &= 5x^2 + 1\end{aligned}$$

For convenience, we often write subtraction of polynomials in vertical form. Thus

$$\begin{array}{r}4x^2 - 2x + 3 \\ -x^2 - 2x + 2 \\ \hline 5x^2 + 1\end{array}$$

Here we simply think of adding the additive inverse of $-x^2 - 2x + 2$, which is $x^2 + 2x - 2$, giving

$$\begin{array}{r}4x^2 - 2x + 3 \\ x^2 + 2x - 2 \\ \hline 5x^2 + 1\end{array}$$

EXERCISES 3.2

Add (Exercises 1–10).

1. $2x + 3$
   $3x - 2$

2. $3x^2 - 4x + 3$
   $2x^2 + 5x - 7$

3. $x^3 - 1$
   $x^3 - 2x + 1$

4. $a^2 + b^2 - c^2$
   $ b^2 + c^2$

5. $x^2y^2 - 2xy + 1$
   $x^2y^2 + 3xy - 4$

6. $az^3 - 2az^2 + 3z$
   $-az^3 + 2z$

7. $a^3 - b^3$
   $a^3 + 3a^2b + 2ab^2 + b^3$

8. $x^2 + 3x - 7$
   $x^2 - 2x + 9$

9. $ab - b^3$
   $-ab - b^3$

10. $ax^2 + by^2 - x$
    $-ax^2 - 4x - 4$

Add (Exercises 11–15).

11. $x + 3x$
12. $(x^2 + 3x) + (4x^2 - 5x)$
13. $(x^2 - x + 1) + x$
14. $(y^2 + 3y + 2) + (y^2 + 2y + 4)$
15. $(b^3 - 1) + (b^3 + 1)$

Subtract (Exercises 16–25).

16. $3x + 4$
    $\underline{4x - 5}$

17. $3x^2 - 2x + 7$
    $\underline{5x^2 + 2x - 7}$

18. $1 - b^2 + b^3$
    $\underline{2 - b^2 + 3b^3}$

19. $x^3 - y^3$
    $\underline{x^3 \quad\quad - 2}$

20. $a - b$
    $\underline{a - b}$

21. $4 - 4bc$
    $\underline{2 - 2bc}$

22. $2x^2 - 3x + 7$
    $\underline{5x^2 + 8x - 9}$

23. $2x^2 - 4x$
    $\underline{x^2 \quad\quad - 7}$

24. $x^2 - 5xy + y^2$
    $\underline{3x^2 + 2xy - y^2}$

25. $x^3 \quad\quad\quad\quad - y^3$
    $\underline{x^3 - 3x^2y + 3xy^2 - y^3}$

Subtract (Exercises 26–30).

26. $(6x^2 - 3x + 2) - (2x^2 - 2x + 7)$
27. $(t^2 - 1) - (t - 1)$
28. $(4y + 5) - (-y - 3)$
29. $(2p^2 + 3p) - (4p^2 - 2p + 1)$
30. $(x^3 - y^3) - (3y^3 - 4)$

3.4 ◆ Multiplication of Polynomials

The process of multiplication of polynomials is based on the commutative and associative properties and on the distributive property. The following examples illustrate the steps in the multiplication of polynomials.

EXAMPLE 1

$(2x)(3x) = 2[x(3x)]$ associative property of multiplication
$ = 2(x[(x)(3)])$ commutative property of multiplication
$ = 2[(x^2)(3)]$ associative property of multiplication
$ = 2[3x^2]$ commutative property of multiplication
$ = [(2)(3)]x^2$ associative property of multiplication
$ = 6x^2$

3.4 MULTIPLICATION OF POLYNOMIALS

EXAMPLE 2

$(x + 1)(x + 2) = (x + 1)(x) + (x + 1)(2)$ distributive property
$= (x^2 + x) + (2x + 2)$ distributive property and the commutative property of multiplication
$= x^2 + (x + 2x) + 2$ associative property of addition, used twice
$= x^2 + x(1 + 2) + 2$ distributive property
$= x^2 + 3x + 2$ commutative property of multiplication

EXAMPLE 3

$(2x^2 + 1)(x^3 + x + 1)$
$= (2x^2 + 1)(x^3) + (2x^2 + 1)(x) + (2x^2 + 1)(1)$ distributive property
$= (2x^5 + x^3) + (2x^3 + x) + (2x^2 + 1)$ distributive property
$= 2x^5 + (x^3 + 2x^3) + 2x^2 + x + 1$ associative and commutative properties of addition
$= 2x^5 + x^3(1 + 2) + 2x^2 + x + 1$ distributive property
$= 2x^5 + 3x^3 + 2x^2 + x + 1$ commutative property of multiplication

When we multiply polynomials, we often use the vertical form whereby we apply the distributive property in the usual way and arrange the terms of the products that are **similar** (that contain the same power of the same base). The product is then easy to find. Thus

$$\begin{array}{r} x^3 + 2x^2 + x - 1 \\ 2x^2 + 3 \\ \hline 2x^5 + 4x^4 + 2x^3 - 2x^2 \\ 3x^3 + 6x^2 + 3x - 3 \\ \hline 2x^5 + 4x^4 + 5x^3 + 4x^2 + 3x - 3 \end{array}$$

EXERCISES 3.3

Multiply (Exercises 1–10).

1. $(2x)(3y)$
2. $(-3x)(-4y)$
3. $(6x)(\frac{1}{2}x^2)(3y)$
4. $(-2x)(-\frac{3}{4}y)(-2xy)$
5. $(\frac{1}{4}rst)(-\frac{2}{3}r^2s^2)$
6. $2x(x + 3)$
7. $3y(2y^2 - 2)$
8. $xy(3y + 4)$
9. $-2y^2(x^2 + y^2)$
10. $-4yz(3x - 2y)$

54 POLYNOMIALS

Multiply (Exercises 11–20).

11. $3x + 7$
    $\underline{2x - 1}$

12. $x + y$
    $\underline{x - y}$

13. $2x + y$
    $\underline{2x + y}$

14. $x^2 - y^2$
    $\underline{x - y}$

15. $a + b$
    $\underline{a - b}$

16. $x^3 - xy + y^2$
    $\underline{x + y}$

17. $x^3 + y^3$
    $\underline{x^3 - y^3}$

18. $x^2 - 2x + 1$
    $\underline{x^2 + 2x - 1}$

19. $a^3 + 3a^2b + 3ab^2 + b^3$
    $\underline{a - b}$

20. $m^2 + m + 3$
    $\underline{m^2 + 2m + 4}$

3.5 ◆ Special Products

In Section 3.4 we found that to multiply two polynomials we use the associative and commutative properties and the distributive property. There are certain pairs of polynomials, however, that we should be able to recognize on sight and give the products instantly.

The first is the product of the sum and difference of the same two terms. Observe the products of the following binomials of this type.

$$(a + b)(a - b) = (a + b)(a) + (a + b)(-b)$$
$$= a^2 + ab - ab + b^2$$
$$= a^2 - b^2$$
$$(2x + 3)(2x - 3) = (2x + 3)(2x) + (2x + 3)(-3)$$
$$= 4x^2 + 6x - 6x - 9$$
$$= 4x^2 - 9$$
$$(3x^2 + 2y)(3x^2 - 2y) = (3x^2 + 2y)(3x^2) + (3x^2 + 2y)(-2y)$$
$$= 9x^4 + 6x^2y - 6x^2y - 4y^2$$
$$= 9x^4 - 4y^2$$

We see that the product of the sum and difference of the same two terms is the difference of their squares:

$$(x + y)(x - y) = x^2 - y^2$$

Another special case is the square of a binomial. Observe the following.

$$(a + b)^2 = (a + b)(a + b)$$
$$= (a + b)(a) + (a + b)(b)$$
$$= a^2 + ab + ab + b^2$$
$$= a^2 + 2ab + b^2$$

3.5 SPECIAL PRODUCTS

$$(x + 2)^2 = (x + 2)(x + 2)$$
$$= (x + 2)(x) + (x + 2)(2)$$
$$= x^2 + 2x + 2x + 4$$
$$= x^2 + 4x + 4$$

We see that when we square a binomial of the form $(a + b)$ the product may be expressed as the sum of three terms; two of the terms, a^2 and b^2, correspond to the squares of the terms of the binomial and the remaining term, $2ab$, is twice the product of the terms of the binomial. Then

$$(a + b)^2 = a^2 + 2ab + b^2$$

Since $(a - b) = [a + (-b)]$, it follows that

$$(a - b)^2 = [a + (-b)]^2$$
$$= a^2 + 2a(-b) + (-b)^2$$
$$= a^2 - 2ab + b^2$$

Now let us consider a product of the form $(x + a)(x + b)$.

$$(x + a)(x + b) = (x + a)(x) + (x + a)(b)$$
$$= x^2 + ax + bx + ab$$
$$= x^2 + (a + b)x + ab$$
$$(x + 3)(x - 2) = (x + 3)(x) + (x + 3)(-2)$$
$$= x^2 + 3x - 2x + (3)(-2)$$
$$= x^2 + (3 - 2)x + (3)(-2)$$
$$= x^2 + x - 6$$
$$(a - 3)(a - 2) = (a - 3)(a) + (a - 3)(-2)$$
$$= a^2 - 3a - 2a + (-3)(-2)$$
$$= a^2 + (-3 - 2)a + (-3)(-2)$$
$$= a^2 - 5a + 6$$

We see from the above examples that the product of two binomials of the form $(x + a)$ and $(x + b)$ is

$$(x + a)(x + b) = x^2 + (a + b)x + ab$$

Observe the products shown below.

$$(ax + by)(cx + dy) = (ax + by)(cx) + (ax + by)(dy)$$
$$= acx^2 + bcxy + adxy + bdy^2$$
$$= acx^2 + (bc + ad)xy + bdy^2$$
$$(5x + 3y)(2x - 3y) = (5x + 3y)(2x) + (5x + 3y)(-3y)$$
$$= 10x^2 + 6xy - 15xy - 9y^2$$
$$= 10x^2 + (6 - 15)xy - 9y^2$$
$$= 10x^2 - 9xy - 9y^2$$

POLYNOMIALS

$$(2x - 3y)(3x - 4y) = (2x - 3y)(3x) + (2x - 3y)(-4y)$$
$$= 6x^2 - 9xy - 8xy + 12y^2$$
$$= 6x^2 + (-9 - 8)xy + 12y^2$$
$$= 6x^2 - 17xy + 12y^2$$

We see from the above that the product of two binomials of the form $(ax + by)$ and $(cx + dy)$ is

$$(ax + by)(cx + dy) = acx^2 + (ad + bc)xy + bdy^2$$

For convenience we restate the five special products discussed above.

$$(a + b)(a - b) = a^2 - b^2$$
$$(a + b)^2 = a^2 + 2ab + b^2$$
$$(a - b)^2 = a^2 - 2ab + b^2$$
$$(x + a)(x + b) = x^2 + (a + b)x + ab$$
$$(ax + by)(cx + dy) = acx^2 + (ad + bc)xy + bdy^2$$

EXERCISES 3.4

Write the products without doing the computation.

1. $(a - b)(a + b)$
2. $(y + z)^2$
3. $(z - 2)^2$
4. $(2x - y)(2x + y)$
5. $(-y + 4)(-y - 4)$
6. $(2x + 3)^2$
7. $(x + 3)(x + 4)$
8. $(y - 2)(y + 3)$
9. $(t - 3)(t - 4)$
10. $(s + 4)(s - 3)$
11. $(a^2 - 3b^2)^2$
12. $(3m + 2)^2$
13. $(5r - 3)^2$
14. $(5x - 1)(2x + 3)$
15. $(3x + 4y)^2$
16. $(2x + 3)(4x - 5)$
17. $(5 - 2x)(3 - 4x)$
18. $(x^4 + 3y^4)^2$
19. $(x + 10)(x + 5)$
20. $(-2x - 3)(-3x - 5)$
21. $(x - 4y)(7x + 2y)$
22. $(7x - 3y)^2$
23. $(9x - 4y)(9x + 4y)$
24. $(3x + 7y)(-4x - 6y)$
25. $(8a - 7b)^2$
26. $(4xy - 3)(2xy + 7)$
27. $(2x + a)(5x + 3a)$
28. $(r + s)(r - 2s)$
29. $(r^2 - s^2)(r^2 + s^2)$
30. $(9xy + 10)^2$
31. $(4c - 3d)^2$
32. $(12ab - 13)^2$
33. $(7x^2y^2a - 8)(7x^2y^2a + 8)$
34. $(5x - 8y)(x + 6y)$
35. $(-3x + 5)(-3x - 7)$
36. $(3a + 4b)(-2a + 3b)$
37. $(8x^2 - 4y^2)(3x^2 + 7y^2)$
38. $(21)^2 = (20 + 1)^2$
39. $(37)^2 = (40 - 3)^2$
40. $(45)(55) = (50 + 5)(50 - 5)$

3.6 • Division of Polynomials

The definition of **division** of polynomials is similar to the definition of division of real numbers. If P, Q, and R are polynomials and, also, $Q \neq 0$, we have

$$\frac{P}{Q} = R \text{ if and only if } P = QR$$

If a and b are real numbers and $b \neq 0$, we know that the quotient $\frac{a}{b}$ will exist and will be that real number c such that $a = bc$. If, in addition, we know that the real numbers a and b are integers—again with $b \neq 0$—then the quotient $\frac{a}{b}$ will be a rational number, but it need not be an integer. For example, $\frac{4}{2}$ is the integer 2, but $\frac{4}{3}$ is a rational number, not an integer. In the same way, the quotient of two polynomials need not be a polynomial, but it may be. Consider the following examples.

EXAMPLE 1. $\dfrac{24x^3y}{4x^2} = 6xy$, since $(4x^2)(6xy) = 24x^3y$.

EXAMPLE 2. $\dfrac{-32x^2y^2z}{8x^2z} = -4y^2$, since $(8x^2z)(-4y^2) = -32x^2y^2z$.

EXAMPLE 3

$$\frac{12x^2 + 8x^2y + 4y^2}{4y} = \frac{3x^2}{y} + 2x^2 + y$$

since

$$\left[\frac{3x^2}{y} + 2x^2 + y\right](4y) = 12x^2 + 8x^2y + 4y^2$$

Notice that in Examples 1 and 2, the quotient of the polynomials is a polynomial, but in Example 3, the quotient is not a polynomial. At this time, we will restrict ourselves to dividing polynomials in those cases in which the divisor is a monomial. Such division involves the distributive property and the fact that, if a and b are any real numbers with $b \neq 0$, then

$$\frac{a}{b} = a\left(\frac{1}{b}\right) = \left(\frac{1}{b}\right)a$$

EXAMPLE 4

$$\frac{12x^3y^2 + 8x^2y + 4y^2}{4y} = \frac{1}{4y}(12x^3y^2 + 8x^2y + 4y^2)$$

$$= \frac{12x^3y^2}{4y} + \frac{8x^2y}{4y} + \frac{4y^2}{4y}$$

$$= 3x^3y + 2x^2 + y$$

EXAMPLE 5

$$\frac{125z^3y + 50z^2y^2 + 75zy}{5zy} = \frac{1}{5zy}(125z^3y + 50z^2y^2 + 75zy)$$

$$= \frac{125z^3y}{5zy} + \frac{50z^2y^2}{5zy} + \frac{75zy}{5zy}$$

$$= 25z^2 + 10zy + 15$$

EXAMPLE 6

$$\frac{x^m + x^{m-1} + x^{m-2}}{x^n} = \frac{1}{x^n}(x^m + x^{m-1} + x^{m-2})$$

$$= \frac{x^m}{x^n} + \frac{x^{m-1}}{x^n} + \frac{x^{m-2}}{x^n}$$

$$= x^{m-n} + x^{m-n-1} + x^{m-n-2}$$

EXERCISES 3.5

Find the quotients. Assume that all divisors are nonzero.

1. $\dfrac{32x^3yz}{8xyz}$

2. $\dfrac{-28abc}{7ab}$

3. $\dfrac{-x^2y^2}{xy}$

4. $\dfrac{81a^3b^3}{-9ab^2}$

5. $\dfrac{64t^3}{-4t}$

6. $\dfrac{-120r^3s^2}{-10r^2s^2}$

7. $\dfrac{0}{2ab}$

8. $\dfrac{x^{2m}}{x^m}$

9. $\dfrac{x^{2n}y^{3n}}{x^n y^{2n}}$

10. $\dfrac{a^{3n}b^{5n+1}}{a^{2n}b^{5n}}$

11. $\dfrac{6x + 12}{2}$

12. $\dfrac{5x^3 - 10x^2 + 25x}{5x}$

13. $\dfrac{8x^3y^3 - 12x^2y^2}{-4x^2y}$

14. $\dfrac{24x^5 - 36x^4 + 18x^2}{-6x}$

3.7 FACTORING

15. $\dfrac{5x + 10}{5}$

16. $\dfrac{12x^3 - 6x^2 + 9x}{3x}$

17. $\dfrac{25x^3y^3 - 15x^2y^2}{5xy}$

18. $\dfrac{x^6y^2 + x^5y^3 - x^4y^2}{-x^2y^2}$

19. $\dfrac{2a^4b^4 + 4a^3b + 4ab^2}{2ab}$

20. $\dfrac{4x^2y^2 + 8x^3y^3 - 12x^5y^7}{4x^2y^2}$

21. $\dfrac{36x^3y^3 - 45x^2y^2 + 72x^5y^4}{-9x^2y}$

22. $\dfrac{72abc - 64a^2b^2 - 48ab^3}{8ab}$

23. $\dfrac{54a^3b^3 - 48a^2b^2 + 66ab}{-6ab}$

24. $\dfrac{56t^3 + 48t^2 - 80t}{-8t}$

25. $\dfrac{a^{2n} + 2a^n}{a^n}$

26. $\dfrac{a^{n+1} - a^{n+2+1}}{a^{n-1}}$

27. $\dfrac{3m^2n - 6mn^2 + 15mn}{3mn}$

28. $\dfrac{-a^2 + ab}{-a}$

3.7 ◆ Factoring

When we multiplied polynomials, we used the commutative and associative properties of addition and multiplication and the distributive property. In general, if we are given two polynomials P and Q, we can multiply to find the polynomial $R = PQ$. Whenever P, Q, and R are polynomials and $PQ = R$, we call P and Q **factors** of R and say that PQ is a **factorization** of the polynomial R into polynomial factors.

Now given the polynomial R, we wish to find polynomials P and Q such that $R = PQ$. That is, we wish to find the factors of R. In factoring natural numbers we say that a natural number is **factored completely** if each of the factors is a prime number. We now define what we mean when we say that a polynomial is factored completely. We say that a polynomial *over the integers is factored completely* if:

1. It is written as the product of two or more polynomials over the integers.
2. No factor in the factorization can be factored again as the product of two or more polynomials over the integers *which have degree greater than zero*.

Some factorizations of the polynomial $16x^2 - 16y^2$ are

$$\begin{aligned}16x^2 - 16y^2 &= (4x + 4y)(4x - 4y) \\ &= 4(x + y)(4x - 4y) \\ &= 4(x + y)(4)(x - y) \\ &= 16(x + y)(x - y)\end{aligned}$$

POLYNOMIALS

The first factorization given above is not a complete factorization, since the first factor, $4x + 4y$, can be factored into $4(x + y)$, and $x + y$ has degree greater than zero. The second factorization given above is not a complete factorization, since the second factor, $4x - 4y$, can be factored into $4(x - y)$, and $x - y$ has degree greater than zero. Both the third and fourth factorizations above could be considered complete, but we agree to combine all purely numerical factors (that is, polynomials of zero degree) into one integer coefficient, so that we consider $16(x + y)(x - y)$ as a complete factorization of $16x^2 - 16y^2$.

The restrictions given in the definition of the complete factorization of a polynomial over the integers prohibit such factorizations as

$$x + 4 = 4(\tfrac{1}{4}x + 1)$$

since $\tfrac{1}{4}x + 1$ is not a polynomial over the integers. **The generalized distributive property:**

$$ax + ay + az + at + \cdots = a(x + y + z + t + \cdots)$$

is the basic tool used in discovering common factors in the terms of a polynomial. The examples below illustrate the procedure used:

$$8x + 16 = (8)(x) + (8)(2) = 8(x + 2)$$
$$2x^2 + 6x + 8xy = (2x)(x) + (2x)(3) + (2x)(4y)$$
$$= (2x)(x + 3 + 4y)$$
$$x^{3n} + x^{2n} - x^n = (x^n)(x^{2n}) + (x^n)(x^n) - (x^n)(1)$$
$$= (x^n)(x^{2n} + x^n - 1)$$

Many polynomials that contain four or more terms can be factored by applying the distributive property. The following examples illustrate the procedure used.

EXAMPLE 1

$$xy - y + 3x - 3 = (xy - y) + (3x - 3)$$
$$= (y)(x - 1) + (3)(x - 1)$$
$$= (y + 3)(x - 1)$$

EXAMPLE 2

$$am - an + bm - bn = (am - an) + (bm - bn)$$
$$= a(m - n) + b(m - n)$$
$$= (a + b)(m - n)$$

3.7 FACTORING

EXAMPLE 3

$$x^2 - 2xy - x + 2y = (x^2 - 2xy) - (x - 2y)$$
$$= x(x - 2y) - (1)(x - 2y)$$
$$= (x - 1)(x - 2y)$$

Notice that the first step in each of the above examples consists of grouping the terms in pairs. This technique of factoring polynomials is called **grouping by pairs**. We must keep in mind when using the method of grouping by pairs that the ultimate purpose of grouping is to determine the common factors and then to apply the distributive property.

EXERCISES 3.6

Factor completely (Exercises 1–30).

1. $y^3 + 3y$
2. $ax^2 - a^2x$
3. $2y^3 - 6y^3x$
4. $3x - 9ab + 6cx$
5. $5abc + 15bc^2 - 10ac^2$
6. $x^4y^2 - x^3y$
7. $-16ab^3 + 4a^2b^2$
8. $8a^3 - 16a^2$
9. $ab - a^2 + 2ax$
10. $4abc - 8a^2b^2c^2 + 12a^3b^3c^3$
11. $3a - 6a^2b + 9a^3$
12. $2t^2 + 8t - 6$
13. $3f - 12f^3 + 18f^4$
14. $5m^2 - 15m + 20$
15. $4k^2 + 20k - 24$
16. $48x^3 - 96x - 96$
17. $4x^3 - 8x^4y - 6x^2y^2$
18. $3xy - 3x^2y + 6xy^2$
19. $r^2x^2 + 3r^3$
20. $8x^3 - 4a^2x - 72$
21. $-12p^3 + 16p^2t - 48t^3$
22. $15x^5 - 9x^3 + 18x$
23. $ab^2 - 4a^2b^2 - a^3$
24. $5s^2t^2 - 15st^3 + 25st$
25. $-16x^5 - 16x^3y^4 - 16x^4$
26. $18x^3 - 9x^2y - 81x^2y^2$
27. $125a^3b^3 - 25a^2b^2 - 200ab$
28. $4g^3h^3 - 8g^2h + 36g^4h^2$
29. $8ab - 4a^2b^2 - 48ab^3$
30. $108z^4 - 96z^3y + 72z^2y^2$

Factor completely by the "grouping by pairs" method (Exercises 31–50).

31. $ax + ay + bx + by$
32. $xy - xz + ty - tz$
33. $xy - y + 3x - 3$
34. $2a - 2b + ax - bx$
35. $ab - 3a - bx + 3x$
36. $3x + 3y - ax - ay$
37. $xy + xz - 3ty - 3tz$
38. $x^2 - 4x + 6xy - 24y$
39. $5x + 3xy + 3y + 5$
40. $x - y + x^2 - xy$
41. $y^3 + y^2 - 5y - 5$
42. $4ab - 4ac + 2fc - 2fb$
43. $6mp - 2mq^2 + 9np - 3nq^2$
44. $6ax - 6ay + x - y$
45. $a^3 - 5a + 3a^2 - 15$
46. $2mp + 4mp - 3np - 6np$
47. $12ab + 4ac - 15db - 5cd$
48. $(a - b)^2 + (a - b)$
49. $(x - y)^2 - (x - y)$
50. $(x + 2y)^3 - (x + 2y)^2$

3.8 ◆ Factoring Binomials and Trinomials

In the section of special products (Section 3.5) we learned that

$$(a + b)(a - b) = a^2 - b^2$$

We refer to a binomial that may be written in the form $a^2 - b^2$ as the **difference of two squares.** Each of the following binomials is the difference of two squares:

$$9x^2 - 16y^2 = (3x)^2 - (4y)^2$$
$$64a^4 - 1 = (8a^2)^2 - 1^2$$
$$(x^2 + 2xy + y^2) - (a^2 - 4a + 4) = (x + y)^2 - (a - 2)^2$$

It is quite easy to factor the difference of two squares. Since $(a+b)(a-b) = a^2 - b^2$, we see that the difference of two squares can be factored into the sum and difference of their square roots.

EXAMPLE 1. Factor completely $x^2 - 16$.

SOLUTION. $x^2 - 16 = x^2 - 4^2$
$= (x - 4)(x + 4)$

EXAMPLE 2. Factor completely $4y^2 - 25$.

SOLUTION. $4y^2 - 25 = (2y)^2 - 5^2$
$= (2y + 5)(2y - 5)$

Certain trinomials are factorable as the square of a binomial. Such polynomials are called **perfect squares.** Notice that

$$(x + y)^2 = (x + y)(x + y)$$
$$= x^2 + 2xy + y^2$$

We see that a trinomial is a perfect square whenever:

1. Two of its terms are perfect squares.
2. The third term is twice the product of the square roots of the two terms that are perfect squares.

For example,

$$x^2 + 6x + 9 = x^2 + (2)(3)(x) + 3^2 = (x + 3)^2$$
$$4x^2 - 12x + 9 = (2x)^2 + (2)(2x)(-3) + (-3)^2$$
$$= [2x + (-3)]^2 = (2x - 3)^2$$

3.8 FACTORING BINOMIALS AND TRINOMIALS

Although it is easy to find the product of two binomials, for example, $(x + 3)(x + 4) = x^2 + 7x + 12$, it is not always easy to reverse the process to find the factorization of their product. The problem is simplified by recognizing certain patterns for multiplying binomials. Observe the example below.

$$\begin{aligned}(x + 3)(x + 4) &= (x + 3)(x) + (x + 3)(4) \\ &= x^2 + 3x + 4x + (3)(4) \\ &= x^2 + (3 + 4)x + 12 \\ &= x^2 + 7x + 12\end{aligned}$$

Notice that the first term of the product, x^2, is the product of the first terms of the binomial factors; and the last term of the product, 12, is the product of the last terms of the binomial factors. Finally, notice that the middle term of the product, $7x$, is the sum of the products $3x$ and $4x$.

Now let us see how this example aids us in factoring the trinomial $x^2 - 5x + 6$. We see that the product of the first terms of the binomial factors is x^2, hence the first term in each factor is x. The product of the second terms of the factors is 6, hence the second terms have the same signs and are either 1 and 6, 2 and 3, -1 and -6, or -2 and -3. Since the middle term of the product is $-5x$, the sum of the second terms of the factors must be -5. Since $(-2) + (-3) = -5$, we have

$$x^2 - 5x + 6 = (x - 2)(x - 3)$$

EXAMPLE 3. Factor completely $x^2 + x - 2$.

SOLUTION. The first term of each of the factors is x, since $x \cdot x = x^2$. The product of the second terms of the factors is -2, hence the second terms of the factors have opposite signs, and are either -1 and 2 or 1 and -2. Since the sum of the second terms of the factors is 1, and $2 + (-1) = 1$, we have

$$x^2 + x - 2 = (x + 2)(x - 1)$$

The most difficult trinomial to factor is that occurring in the product

$$(ax + b)(cx + d) = acx^2 + (ad + bc)x + bd$$

We see that we are attempting to factor a trinomial of the form $px^2 + qx + r$ into two factors of the form $(ax + b)(cx + d)$. From the above we see that

$$px^2 + qx + r = acx^2 + (ad + bc)x + bd$$

We observe that we must find four integers, a, b, c, and d, such that

$$ac = p$$
$$bd = r$$
and
$$ad + bc = q$$

Let us try to factor the trinomial

$$3x^2 + 2x - 5$$

It should be clear that only the positive factors of 3 need be considered. The only positive factors of 3 are 1 and 3. Since $r = bd = -5$, we see that b and d have opposite signs and are factors of -5. The possible factors of -5 are 5 and -1 and -5 and 1. *Some* of the possible combinations are shown in the table below.

$p = ac = 3$		$r = bd = -5$		$q = ad + bc = 2$
a	c	b	d	$ad + bd$
3	1	-1	5	$15 - 1 = 14 \neq 2$
3	1	1	-5	$-15 + 1 = -14 \neq 2$
3	1	5	-1	$-3 + 5 = 2$

We see from the above that

$$3x^2 + 2x - 5 = (3x + 5)(x - 1)$$

The above process is usually accomplished mentally.

EXAMPLE 4. Factor completely $8x^2 + 2x - 15$.

SOLUTION. We see that $ac = 8$, $bd = -15$ and $ad + bc = 2$. The possible factors of 8 are 1 and 8 and 2 and 4. The possible factors of -15 are -1 and 15, 15 and -1, 3 and -5, and -3 and 5. Trying the possible combinations, we find that

$$ad + bc = (4)(3) + (-5)(2) = 12 - 10 = 2$$

and hence

$$8x^2 + 2x - 15 = (4x - 5)(2x + 3)$$

If a quadratic trinomial has a common monomial factor, the factorization will be simplified by first factoring this common monomial factor before seeking other factors. For example,

3.8 FACTORING BINOMIALS AND TRINOMIALS

$$12x^2 - 2x - 24 = 2(6x^2 - x - 12)$$
$$= 2(2x - 3)(3x + 4)$$

Finally, it should be noted that for some quadratic trinomials, an attempt to find first degree binomial factors that are polynomials over the integers may not be successful for the simple reason that there are no such factors of that particular trinomial. For example,

$$x^2 + x + 1$$

has no polynomials factors that are polynomials over the integers other than itself and 1.

EXERCISES 3.7

Factor completely.

1. $x^2 + 6x + 9$
2. $x^2 + 8x + 16$
3. $y^2 - 10y + 25$
4. $4x^2 + 4x + 1$
5. $x^2 - 2x + 1$
6. $25 + 10m + m^2$
7. $4x^2 - 12xy + 9y^2$
8. $16a^2 - b^2$
9. $9x^2 - 16$
10. $1 - 169y^2$
11. $100 - x^2$
12. $81y^2 - 4$
13. $x^2 - 25$
14. $4x^4 - 1$
15. $25 - 9y^2$
16. $x^2 + 7x + 12$
17. $x^2 - 5x + 6$
18. $x^2 - x - 90$
19. $x^2 + 3x - 28$
20. $x^2 - 4x - 21$
21. $x^2 - 11x + 28$
22. $x^2 - x - 56$
23. $y^2 - y - 12$
24. $x^2 - 10x + 16$
25. $x^2 - 3x - 4$
26. $x^2 + 9x - 10$
27. $x^2 + x - 20$
28. $3x^2 + 5x + 2$
29. $3x^2 + 10x + 3$
30. $2y^2 + 5y + 3$
31. $2r^2 - 7r + 5$
32. $2k^2 - 7k + 6$
33. $3x^2 + 17x + 10$
34. $3y^2 - 4y - 4$
35. $6x^2 + x - 5$
36. $3x^2 - 2x - 5$
37. $12x^2 - 17x + 6$
38. $6x^2 - 13x - 5$
39. $3y^2 - 17y - 28$
40. $5x^2 + 11x + 2$
41. $3x^2 + 8x + 5$
42. $5x^2 + 14x - 3$
43. $2x^2 + 3x - 2$
44. $9x^4 - 68x^2y^2 - 32y^4$
45. $16y^6 - 26y^3z + 9z^2$
46. $15x^2 + xy - 2y^2$
47. $a^{2n} - b^{2n}$
48. $y^4 - x^4$
49. $3(x + y)^2 + 7(x + y) + 2$
50. $16x^4 - y^4$

3.9 ♦ Factoring the Sum and Difference of Two Cubes

We see that
$$a^3 + b^3 = (a + b)(a^2 - ab + b^2)$$
since
$$\begin{aligned}(a + b)(a^2 - ab + b^2) &= (a + b)a^2 + (a + b)(-ab) + (a + b)b^2 \\ &= a^3 + a^2b - a^2b - ab^2 + ab^2 + b^3 \\ &= a^3 + b^3\end{aligned}$$

It is also true that
$$a^3 - b^3 = (a - b)(a^2 + ab + b^2)$$

The reader should verify the truth of this statement.

By observing the two products above, we have a method for factoring the sum of two cubes and the difference of two cubes. Notice that the first factor of the sum of two cubes is the sum of their cube roots; and the second factor is the sum of the square of the cube root of the first, the opposite of the product of the two cube roots and the square of the cube root of the second.

EXAMPLE 1. Factor completely $x^3 + 125$.

SOLUTION. $$\begin{aligned}x^3 + 125 &= (x)^3 + (5)^3 \\ &= (x + 5)(x^2 + [-5x] + 5^2) \\ &= (x + 5)(x^2 - 5x + 25)\end{aligned}$$

EXAMPLE 2. Factor completely $8y^3 + 27$.

SOLUTION. $$\begin{aligned}8y^3 + 27 &= (2y)^3 + (3)^3 \\ &= (2y + 3)[(2y)^2 + (-(2y)(3)) + (3)^2] \\ &= (2y + 3)(4y^2 - 6y + 9)\end{aligned}$$

In factoring the difference of two cubes we notice that the first factor is the difference of the cube root of the first and second terms. The second factor is the sum of the square of the cube root of the first term, the product of their cube roots and the square of the cube root of the second term.

EXAMPLE 3. Factor completely $64a^3 - 1$.

CHAPTER 3 REVIEW

SOLUTION. $64a^3 - 1 = (4a)^3 - (1)^3$
$= (4a - 1)[(4a)^2 + (4a)(1) + (1)^2]$
$= (4a - 1)(16a^2 + 4a + 1)$

EXAMPLE 4. Factor completely $343p^3 - 125q^6$.

SOLUTION. $343p^3 - 125q^6 = (7p)^3 - (5q^2)^3$
$= (7p - 5q^2)[(7p)^2 + (7p)(5q^2) + (5q^2)^2]$
$= (7p - 5q^2)(49p^2 + 35pq^2 + 25q^4)$

EXERCISES 3.8

Factor completely.

1. $x^3 - y^3$
2. $c^3 + d^3$
3. $y^3 - 8$
4. $x^3 + 1$
5. $x^9 - y^{18}$
6. $x^3 - 64$
7. $a^{12} - b^6$
8. $y^3 + 27$
9. $8x^3y^9 + 64z^3$
10. $8x^3 + 27y^3$
11. $a^{12} + 216b^6$
12. $64x^3 + 27p^3$
13. $343 + 8x^3$
14. $(x + y)^3 - (z + w)^3$
15. $(x + y)^3 + 27z^3$
16. $(a + b)^3 - (a - b)^3$
17. $(5a + b)^3 - (c + d)^3$
18. $(2a + b)^3 + (t + k)^3$
19. $(n + 1)^3 - (n - 2)^3$
20. $x^6 - 1$

CHAPTER 3 REVIEW

Perform the indicated operations and simplify (Exercises 1–4).

1. $(x^5 - 2x^4 + x^2 - 8) + (2x^6 - 3x^5 + x^3 - 2x + 1)$
2. $(2x^3 - 4x^2 + 7x + 3) - (3x^3 + 5x^2 + 3x - 4)$
3. $(2x^2 - 3x + 1)(4x - 5)$
4. $(x^2 - y^2)(x + 3xy)$

Factor completely (Exercises 5–20).

5. $9x^2 - 4y^2$
6. $16x^4 - y^6$
7. $4x^2 + 20xy + 25y^2$
8. $9x^2 - 12xy + 4y^2$
9. $4x + 2xy - 6y - 12$
10. $x^2 - 3x - 4$
11. $x^2 + 3x + 2$
12. $2x^2 - x - 1$
13. $3x^2 - x - 2$
14. $2x^2 - x - 6$
15. $2x^2 + 5x + 3$
16. $6x^2 - x - 1$
17. $6x^2 + 11x - 10$
18. $x^6 - 64$
19. $8x^3 - 1$
20. $x^3 + 27$

CHAPTER 4

Rational Expressions

4.1 ◆ Definitions

A **rational expression,** sometimes called a **fraction,** is an expression of the form $\dfrac{P}{Q}$, where P and Q are polynomials and Q is *not* the polynomial zero. Every algebraic expression that is the quotient of two polynomials is said to be a **rational expression.** Examples of rational expressions are

$$\frac{3x+2}{x+5}, \qquad \frac{x^2+4}{x^2-4}, \qquad \frac{8x+1}{x}$$

When the variables in a rational expression are replaced by real numbers, the numerator and denominator are real numbers. For any such replacements for which the denominator does not have the value zero, the rational expression represents a real number. We must restrict the replacement set(s) of the variable(s) to exclude values that give the denominator the value zero. When a polynomial has value zero, we say that it **vanishes.** A rational expression is said to be **undefined** for any value or values of the variable(s) for which the denominator vanishes. Thus the rational expression

$$\frac{x^2+8}{x^2-4}$$

is undefined for the values 2 and -2, since $(2)^2 - 4 = 0$ and $(-2)^2 - 4 = 0$.

4.2 ◆ Simplification of Rational Expressions

The procedure used in simplifying rational expressions is based on the following property of real numbers: **IF** a, b **AND** c **ARE REAL NUMBERS,** $b \neq 0$ **AND** $c \neq 0$, **THEN**

$$\frac{ac}{bc} = \frac{a}{b}$$

4.2 SIMPLIFICATION OF RATIONAL EXPRESSIONS

A similar property holds for polynomials: **IF P, Q, AND R ARE POLYNOMIALS AND NEITHER Q NOR R IS THE POLYNOMIAL ZERO, THEN**

$$\frac{PR}{QR} = \frac{P}{Q}$$

Simplifying rational expressions consists of two steps:

1. Factoring the numerator and denominator.
2. Applying the property $\frac{PR}{QR} = \frac{P}{Q}$ for Q not 0 and R not 0.

The examples below illustrate the procedure used in simplifying rational expressions.

EXAMPLE 1

$$\frac{x^2 y^3}{x^3 y^2} = \frac{(x^2 y^2)y}{(x^2 y^2)x} = \frac{y}{x} \qquad x \neq 0 \text{ and } y \neq 0$$

EXAMPLE 2

$$\frac{x^2 - y^2}{2x - 2y} = \frac{(x+y)(x-y)}{2(x-y)} = \frac{x+y}{2} \qquad x \neq y$$

EXAMPLE 3

$$\frac{4x^2 - 12x + 9}{4x^2 - 9} = \frac{(2x-3)(2x-3)}{(2x+3)(2x-3)} = \frac{2x-3}{2x+3} \qquad x \neq \frac{3}{2}, x \neq -\frac{3}{2}$$

EXAMPLE 4

$$\frac{xy + 3x + y^2 - 9}{y^2 + 5y + 6} = \frac{(x+y-3)(y+3)}{(y+2)(y+3)} = \frac{x+y-3}{y+2}$$

$$y \neq -2, y \neq -3$$

EXERCISES 4.1

Specify any real values of the variables for which the following rational expressions are undefined (Exercises 1–8).

1. $\dfrac{3}{x}$

2. $\dfrac{x+4}{x-2}$

3. $\dfrac{y}{y-1}$

4. $\dfrac{3x}{x^2-9}$

RATIONAL EXPRESSIONS

5. $\dfrac{x-4}{x^2-5x+6}$

6. $\dfrac{5x+7}{6x^2-11x-10}$

7. $\dfrac{3x+1}{x^2+4x+4}$

8. $\dfrac{2xy}{x^2-16}$

Simplify. Assume that no divisors are zero (Exercises 9–30).

9. $\dfrac{x^2y}{xy}$

10. $\dfrac{3x^2y^3}{-3y^5}$

11. $\dfrac{3x}{12x^2}$

12. $\dfrac{5y^3}{-15y}$

13. $\dfrac{5(x+y)}{15(x+y)}$

14. $\dfrac{x^2-x^2y}{1-y}$

15. $\dfrac{5x-5}{x^2-1}$

16. $\dfrac{2y^2-2}{y-1}$

17. $\dfrac{x^2+6x+9}{x+3}$

18. $\dfrac{3x^3-24}{3x^2+6x+12}$

19. $\dfrac{x^2+4x+4}{xy-3x+2y-6}$

20. $\dfrac{4x^2-12xy+9y^2}{4x^2-9y^2}$

21. $\dfrac{x^3+8}{x^2-2x-8}$

22. $\dfrac{y^2-27}{y^2+y-12}$

23. $\dfrac{6r^2-5r-6}{6r^2-11r-10}$

24. $\dfrac{b^2-81}{b^2-3b-54}$

25. $\dfrac{x^2-x+1}{x^3+1}$

26. $\dfrac{x^2-x-2}{2x^2-5x+2}$

27. $\dfrac{-(b-a)}{b^2-a^2}$

28. $\dfrac{xy+3x+y^2-9}{x^2+2xy+y^2-9}$

29. $\dfrac{x^3-27}{x^2+3x+9}$

30. $\dfrac{x^4-2x^2y^2+y^4}{x^2+2xy+y^2}$

4.3 ◆ Least Common Multiples

The least common multiple (L.C.M.) of two or more natural numbers is the smallest natural number that is exactly divisible by each of the numbers. For example, the least common multiple of 3 and 5 is 15; the least common multiple of 4 and 6 is 12. To find the L.C.M. of two or more natural numbers:

1. Write each number as a product of primes.
2. Find the product of all the different prime factors occurring in any of the numbers, using each prime factor the greatest number of times it occurs in any one of the numbers.

4.4 ADDITION AND SUBTRACTION OF RATIONAL EXPRESSIONS

For example, the L.C.M. of 12, 16, and 25 is found in the following manner.

1. $12 = (2^2)(3)$; $16 = 2^4$; $15 = (3)(5)$.
2. The L.C.M. is $(2^4)(3)(5) = 240$.

The least common multiple of several polynomials is found in a manner similar to that of finding the L.C.M. of natural numbers. The procedure is as follows.

1. *Write each polynomial as the product of polynomials that are factored completely.*
2. *Find the product of all the different polynomial factors occurring in any of the polynomials, using each polynomial factor the greatest number of times it occurs in any factorization.*

EXAMPLE. Find the L.C.M. of

$$x^2 - y^2, \quad x^2 - 2xy + y^2, \quad \text{and} \quad x^3 - y^3$$

1. $x^2 - y^2 = (x + y)(x - y)$
   $x^2 - 2xy + y^2 = (x - y)^2$
   $x^3 - y^3 = (x - y)(x^2 + xy + y^2)$
2. The L.C.M. is $(x + y)(x - y)^2(x^2 + xy + y^2)$.

EXERCISES 4.2

Find the L.C.M. of each of the following.

1. 8, 12
2. 3, 4, 6
3. 14, 21, 42
4. 16, 36, 42
5. $2ab$, b^2, $6a^3$
6. $7x$, $9y$, $12xy$
7. x^2y^2, x^3y, $3xy^3$
8. ab, ab^2, b, a
9. $x^2 - y^2$, $ax + ay$
10. $x + y$, $x - y$, $x^2 - y^2$
11. $x^2 - 4$, $x + 2$, $x^2 + 5x + 6$
12. $x + 4$, $x^2 + 8x + 16$, $x^2 + 7x + 12$
13. $x^2 - 3x + 2$, $x^2 - 6x + 8$
14. $x^2 + x$, $x + 1$
15. $(x - 2)^2$, $2x - 4$
16. $x^2 - 4$, $x^2 - 10x + 16$
17. $x^2 - 5x + 4$, $(x - 1)^3$
18. $y^3 - 1$, $(y - 1)^3$
19. $x^3 - y^3$, $x + y$, $x^2 + xy + y^2$
20. $(x + y)^3$, $x^2 - y^2$

4.4 ♦ Addition and Subtraction of Rational Expressions

The addition of rational expressions makes use of the distributive property and the following property of real numbers.

RATIONAL EXPRESSIONS

IF a AND b ARE REAL NUMBERS, AND $b \neq 0$, THEN

$$\frac{a}{b} = a\left(\frac{1}{b}\right)$$

For polynomials, we have this property.

IF P, Q, AND R ARE POLYNOMIALS AND Q IS NOT THE POLYNOMIAL ZERO, THEN

$$\frac{P}{Q} + \frac{R}{Q} = P\left(\frac{1}{Q}\right) + R\left(\frac{1}{Q}\right)$$
$$= (P + R)\left(\frac{1}{Q}\right)$$
$$= \frac{P + R}{Q}$$

From the above we see that the sum of two rational expressions having the same denominator is also a rational expression; its numerator is the sum of the numerators, and its denominator is the common denominator.

EXAMPLE 1

$$\frac{6}{x} + \frac{3}{x} = \frac{6 + 3}{x} = \frac{9}{x} \qquad x \neq 0$$

EXAMPLE 2

$$\frac{x + 3}{y} + \frac{x + 4}{y} = \frac{(x + 3) + (x + 4)}{y} = \frac{2x + 7}{y} \qquad y \neq 0$$

EXAMPLE 3

$$\frac{3x - 2}{x^2 + y^2} + \frac{5x + 7}{x^2 + y^2} = \frac{(3x - 2) + (5x + 7)}{x^2 + y^2} = \frac{8x + 5}{x^2 + y^2} \qquad (x^2 + y^2) \neq 0$$

When rational expressions have different denominators, we must find equivalent expressions with a common denominator and then add. The least common denominator (L.C.D.) of two or more rational expressions is the L.C.M. of the denominators of the rational expressions.

To find a rational expression equivalent to a given rational expression, we multiply numerator and denominator of the given rational expression by the same polynomial. We can do this because of the following property of real numbers:

4.4 ADDITION AND SUBTRACTION OF RATIONAL EXPRESSIONS

IF a, b, AND c ARE REAL NUMBERS, $c \neq 0$, $b \neq 0$, THEN

$$\frac{a}{b} = \frac{ac}{bc}$$

Suppose we wish to find rational expressions equivalent to

$$\frac{x+y}{x^2-y^2} \quad \text{and} \quad \frac{2x-y}{x^2+xy-2y^2}$$

having the least common denominator. The least common denominator of the two fractions is the least common multiple of these two denominators. The least common multiple of $x^2 - y^2 = (x+y)(x-y)$ and $x^2 + xy - 2y^2 = (x-y)(x+2y)$ is $(x-y)(x+y)(x+2y)$. We now multiply numerator and denominator of $\dfrac{x+y}{x^2-y^2}$ by $(x+2y)$ and of $\dfrac{2x-y}{x^2+xy-2y^2}$ by $(x+y)$ to obtain

$$\frac{(x+y)(x+2y)}{(x+y)(x-y)(x+2y)} \quad \text{and} \quad \frac{(2x-y)(x+y)}{(x-y)(x+2y)(x+y)}$$

We now have two rational expressions equivalent to the given rational expressions and which have the same denominator. We can now add these two rational expressions by adding their numerators as shown above.

Consider the following examples.

EXAMPLE 1

$$\frac{3y}{y^2-4} + \frac{2}{y+2} = \frac{3y}{(y+2)(y-2)} + \frac{2}{y+2}$$
$$= \frac{3y}{(y+2)(y-2)} + \frac{2(y-2)}{(y+2)(y-2)}$$
$$= \frac{3y+2y-4}{(y+2)(y-2)}$$
$$= \frac{5y-4}{(y+2)(y-2)} \quad y \neq 2, y \neq -2$$

EXAMPLE 2

$$\frac{x-1}{x^2-9} + \frac{x+3}{x^2-6x-9} = \frac{x-1}{(x+3)(x-3)} + \frac{x+3}{(x-3)^2}$$
$$= \frac{(x-1)(x-3)}{(x+3)(x-3)^2} + \frac{(x+3)(x+3)}{(x+3)(x-3)^2}$$
$$= \frac{(x^2-4x+3)+(x^2+6x+9)}{(x+3)(x-3)^2}$$
$$= \frac{2x^2+2x+12}{(x+3)(x-3)^2} \quad x \neq 3, x \neq -3$$

Since the set of polynomials is a subset of the set of rational expressions, we must define the difference of two rational expressions in a way which is consistent with the definition of the difference of two polynomials. For all rational expressions, we have

$$\frac{P}{Q} - \frac{R}{S} = \frac{P}{Q} + \left(-\frac{R}{S}\right)$$

where P, Q, R, and S are polynomials and neither Q nor S is the zero polynomial. Thus the difference, $\frac{P}{Q} - \frac{R}{S}$, is defined as the sum of $\frac{P}{Q}$ and the additive inverse of $\frac{R}{S}$.

The additive inverse of the rational expression $\frac{R}{S}$ can be written in three different ways:

$$-\frac{R}{S}, \quad \frac{-R}{S}, \quad \text{and} \quad \frac{R}{-S}$$

Thus the additive inverse of $\frac{x-1}{x+1}$ is

$$-\frac{x-1}{x+1}, \quad \frac{-(x-1)}{x+1} = \frac{-x+1}{x+1}, \quad \text{and} \quad \frac{x-1}{-(x+1)} = \frac{x-1}{-x-1}$$

The following examples illustrate the procedure for subtracting one rational expression from another.

EXAMPLE 1

$$\frac{3}{x} - \frac{5}{x} = \frac{3}{x} + \frac{-5}{x} = \frac{3+(-5)}{x} = \frac{-2}{x} \quad x \neq 0$$

EXAMPLE 2

$$\frac{y^2}{5} - \frac{y}{5} = \frac{y^2}{5} + \frac{-y}{5} = \frac{y^2 + (-y)}{5} = \frac{y^2 - y}{5}$$

EXAMPLE 3

$$\frac{5}{x-2} - \frac{3}{x^2-4} = \frac{5(x+2)}{(x+2)(x-2)} + \frac{-3}{(x+2)(x-2)}$$

$$= \frac{5(x+2) + (-3)}{(x+2)(x-2)}$$

$$= \frac{5x+7}{(x+2)(x-2)} \quad x \neq 2, x \neq -2$$

4.4 ADDITION AND SUBTRACTION OF RATIONAL EXPRESSIONS

EXAMPLE 4

$$\frac{x+2}{(x-1)(x-2)} - \frac{x+3}{(x-1)(2-x)} = \frac{x+2}{(x-1)(x-2)} + \frac{x+3}{(-1)(x-1)(2-x)}$$

$$= \frac{x+2}{(x-1)(x-2)} + \frac{x+3}{(x-1)(x-2)}$$

$$= \frac{(x+2)+(x+3)}{(x-1)(x-2)}$$

$$= \frac{2x+5}{(x-1)(x-2)} \qquad x \neq 1, x \neq 2$$

The three ways of writing the additive inverse of a rational expression are often referred to as "**the three signs of a fraction,**" which are:

1. A sign for the fraction.
2. A sign for the numerator.
3. A sign for the denominator.

If any two of the three signs are changed, the resulting fraction is equal to the original fraction. That is,

$$\frac{P}{Q} = \frac{-P}{-Q} = -\frac{-P}{Q} = -\frac{P}{-Q}$$

and

$$-\frac{-P}{-Q} = -\frac{P}{Q} = \frac{P}{-Q} = \frac{-P}{Q}$$

If the denominator of the fraction is an expression containing more than one term, by changing two signs of the fraction we obtain

$$-\frac{-a}{a-b} = -\frac{(-1)(-a)}{(-1)(a-b)} = -\frac{a}{b-a}$$

EXERCISES 4.3

Which of the following are true for all values of x? Assume that no denominator has value zero (Exercises 1–10).

1. $-\dfrac{x}{3} = \dfrac{-x}{3}$

2. $\dfrac{x-2}{3} = -\dfrac{2-x}{3}$

3. $-\dfrac{x}{x-2} = \dfrac{-x}{2-x}$

76 RATIONAL EXPRESSIONS

4. $\dfrac{x^2 + 2x + 1}{x} = \dfrac{-x^2 - 2x + 1}{-x}$

5. $\dfrac{x(1 - x)}{3} = \dfrac{-x(x - 1)}{3}$

6. $\dfrac{5}{x} - \dfrac{3}{y} = \dfrac{5}{x} + \dfrac{-3}{y}$

7. $\dfrac{2}{x - 1} - \dfrac{3}{1 - x} + \dfrac{5}{1 - x} = \dfrac{2}{x - 1} + \dfrac{3}{x - 1} + \dfrac{5}{1 - x}$

8. $\dfrac{x - 1}{(x - 2)(x - 3)} - \dfrac{x + 2}{(3 - x)(x - 2)} = \dfrac{x - 1}{(x - 2)(x - 3)} + \dfrac{x + 2}{(x - 3)(x - 2)}$

9. $\dfrac{-5}{-3 - x} = \dfrac{5}{x + 3}$

10. $\dfrac{x}{x - 1} - \dfrac{-3x}{1 - x} = \dfrac{x}{x - 1} - \dfrac{3x}{x - 1}$

Add. Assume that no denominator has a value of zero (Exercises 11–30).

11. $\dfrac{3}{y} + \dfrac{7}{y}$

12. $\dfrac{x}{ab} + \dfrac{2x}{ab}$

13. $\dfrac{2a + b}{a - b} + \dfrac{a + 3b}{a - b}$

14. $\dfrac{5x + y}{x^2 - y^2} + \dfrac{3x - 2y}{x^2 - y^2}$

15. $\dfrac{3a + 4}{a^2 + 2a + 1} + \dfrac{4a - 5}{a^2 + 2a + 1}$

16. $\dfrac{5}{3 - ab} + \dfrac{3}{3 - ab}$

17. $\dfrac{y + 2}{x^2 - x + 2} + \dfrac{-2y - 4}{x^2 - x + 2}$

18. $\dfrac{3}{xy} + \dfrac{4x}{x^2y^2}$

19. $\dfrac{1}{x} + \dfrac{1}{y} + \dfrac{1}{xy}$

20. $\dfrac{x^2}{x^2 + 1} + \dfrac{x^3 - 1}{x^3 + 1}$

21. $\dfrac{x}{ax - x^2} + \dfrac{b}{x^2 - ax}$

22. $\dfrac{2}{5x - 10} + \dfrac{5}{6x - 12}$

23. $\dfrac{1}{y^2 + xy} + \dfrac{1}{xy + x^2}$

24. $\dfrac{2x}{4x^2 + 4x + 1} + \dfrac{-1}{2x + 1}$

25. $\dfrac{a - 1}{a^2 - 9} + \dfrac{a + 2}{a + 3}$

26. $\dfrac{2x}{(x - 2)^2} + \dfrac{4x}{2x - 4}$

27. $\dfrac{x - a}{x + a} + \dfrac{x + a}{x - a}$

28. $\dfrac{x + 2y}{x - y} + \dfrac{x - y}{x + 2y}$

29. $\dfrac{x - a}{x^2 - b^2} + \dfrac{x - b}{x + b}$

30. $\dfrac{3x + 2}{x^2 + 6x + 9} + \dfrac{3}{x + 3}$

Subtract. Assume that no denominator has value zero (Exercises 31–40).

31. $\dfrac{x^2}{y} - \dfrac{3}{y}$

32. $\dfrac{2x}{3y^2} - \dfrac{4x}{3y^2}$

4.5 MULTIPLICATION OF RATIONAL EXPRESSIONS

33. $\dfrac{3x}{y^2} - \dfrac{4y}{y}$

34. $\dfrac{1}{x} - \dfrac{1}{y}$

35. $\dfrac{2x+2}{x^2-1} - \dfrac{x}{x+1}$

36. $\dfrac{y}{y+3} - \dfrac{y-4}{y}$

37. $\dfrac{2}{x^2-4x+4} - \dfrac{-x}{x-2}$

38. $\dfrac{4}{2-x} - \dfrac{6}{-2+x}$

39. $\dfrac{a}{ax-x^2} - \dfrac{x}{a^2-ax}$

40. $\dfrac{y}{y^2-1} - \dfrac{y+1}{y^2-4y+3}$

4.5 ◆ Multiplication of Rational Expressions

We define the product of two rational expressions in a manner which is consistent with the multiplication of rational numbers. Thus, if P, Q, R, and S are polynomials, $Q \neq 0$, and $S \neq 0$, then

$$\frac{P}{Q} \cdot \frac{R}{S} = \frac{PR}{QS}$$

Since polynomials are real numbers when the variables are replaced by real numbers, we can use the properties of the real number system to show the validity of this multiplication. Let

$$\frac{P}{Q} = x \text{ and } \frac{R}{S} = y$$

Then

$P = Qx$ and $R = Sy$ definition of division
$PR = (Qx)(Sy)$ closure property of multiplication
$= (xy)(QS)$ associative and commutative properties of multiplication

so that

$$xy = \frac{PR}{QS} \qquad \text{definition of division}$$

Thus the product of two rational expressions is another rational expression in which the numerator is the product of the numerators and the denominator is the product of the denominators.

In practice, it is usually desirable to factor, if possible, the numerators and denominators of the rational expressions being multiplied. The following examples illustrate the procedure followed when multiplying rational expressions.

RATIONAL EXPRESSIONS

EXAMPLE 1

$$\frac{5}{x} \cdot \frac{3}{y} = \frac{(5)(3)}{(x)(y)} = \frac{15}{xy} \qquad x \neq 0, y \neq 0$$

EXAMPLE 2

$$\frac{ax + ay}{ab} \cdot \frac{b^3}{bx + by} = \frac{a(x + y)}{ab} \cdot \frac{b^3}{b(x + y)}$$

$$= \frac{ab^3(x + y)}{ab^2(x + y)} = b \qquad a \neq 0, b \neq 0, x \neq -y$$

EXAMPLE 3

$$\frac{x^2 - 9}{x + 5} \cdot \frac{x^2 - 25}{3 - x} = \frac{(x + 3)(x - 3)}{x + 5} \cdot \frac{(x + 5)(x - 5)}{(-1)(x - 3)}$$

$$= \frac{(x + 3)(x - 3)(x + 5)(x - 5)}{(-1)(x + 5)(x - 3)}$$

$$= \frac{(x + 3)(x - 5)}{(-1)}$$

$$= -(x + 3)(x - 5) \qquad x \neq 3, x \neq -5$$

EXAMPLE 4

$$\frac{x^3 + 1}{x^2 - 4y^2} \cdot \frac{x - 2y}{x^2 - x + 1} = \frac{(x + 1)(x^2 - x + 1)}{(x + 2y)(x - 2y)} \cdot \frac{x - 2y}{x^2 - x + 1}$$

$$= \frac{(x + 1)(x^2 - x + 1)(x - 2y)}{(x + 2y)(x - 2y)(x^2 - x + 1)}$$

$$= \frac{x + 1}{x + 2y}$$

$$x + 2y \neq 0, x - 2y \neq 0, x^2 - x + 1 \neq 0$$

EXERCISES 4.4

Find the products. Assume that no denominator has zero value.

1. $\dfrac{x}{2} \cdot \dfrac{x^2}{14}$

2. $\dfrac{m}{n} \cdot \dfrac{n^3}{m^2}$

3. $\dfrac{7a}{3b} \cdot \dfrac{9b^3}{21a^2}$

4. $\dfrac{x}{y} \cdot \dfrac{xy}{x - y}$

5. $\dfrac{x^2 - y^2}{xy} \cdot \dfrac{x^2 y^2}{x + y}$

6. $\dfrac{2}{x - 2} \cdot \dfrac{x - 2}{x - 4}$

4.6 DIVISION OF RATIONAL EXPRESSIONS

7. $\dfrac{6y^2 - 24}{y} \cdot \dfrac{4y^3}{2y - 4}$

8. $\dfrac{x + 1}{x - 1} \cdot \dfrac{x^2 - 2x + 1}{x^2 - 1}$

9. $\dfrac{x^2 - 4}{x - 3} \cdot \dfrac{x^2 - x - 6}{x - 2}$

10. $\dfrac{x^2 + 4x}{5x} \cdot \dfrac{30}{x + 4}$

11. $\dfrac{x^2 - 2x - 8}{x - 1} \cdot \dfrac{x^3 - 1}{x^2 + 4x + 4}$

12. $\dfrac{x - 1}{x^2 - 4x} \cdot \dfrac{x^2 - 16}{2}$

13. $\dfrac{x^3 - 8}{a - 2b} \cdot \dfrac{4a - 8b}{x - 2}$

14. $\dfrac{9x^2 - 25}{4x - 4} \cdot \dfrac{x^2 - 1}{3x + 5}$

15. $\dfrac{x^2 - 2x - 8}{x^2 - 25} \cdot \dfrac{10x - 50}{x^2 - 8x + 16}$

16. $\dfrac{x^3 - 27}{x^2 - 2xy} \cdot \dfrac{x^2 - 4y^2}{x^2 + 3x + 9}$

17. $\dfrac{x^2 - 2x - 3}{y - 2x} \cdot \dfrac{y - 2x}{x - 3}$

18. $\dfrac{15xy}{4x^2 - 12x + 9} \cdot \dfrac{2x^2 + 3x - 9}{5y^3}$

19. $\dfrac{x^4 - y^4}{x^2 - 2xy + y^2} \cdot \dfrac{x^2 - y^2}{x^2 + y^2}$

20. $\dfrac{4x^2 - 36}{x^2 + 4x + 3} \cdot \dfrac{x + 1}{6x^2 + 6x - 72}$

4.6 • Division of Rational Expressions

The definition of division of rational numbers must be consistent with the definition of division of real numbers and of polynomials. We know that if a and b are real numbers and $b \neq 0$, then

$$a \div b = \dfrac{a}{b} = c \quad \text{if and only if} \quad a = bc$$

Thus, if P, Q, R, and S are polynomials and none of Q, R, and S is the polynomial zero,

$$\dfrac{P}{Q} \div \dfrac{R}{S} = \dfrac{\frac{P}{Q}}{\frac{R}{S}} = x$$

if and only if

$$\dfrac{P}{Q} = \dfrac{R}{S} \cdot x$$

$$PS = (QR)x$$

$$x = \dfrac{PS}{QR}$$

$$x = \dfrac{P}{Q} \cdot \dfrac{S}{R}$$

We see that, in order to divide one rational expression by another, we must multiply the dividend by the reciprocal of the divisor. The following examples illustrate the procedures.

EXAMPLE 1

$$\frac{x}{\frac{x}{y}} = x \cdot \frac{y}{x}$$

$$= \frac{xy}{x}$$

$$= y \qquad x \neq 0, y \neq 0$$

EXAMPLE 2

$$\frac{\frac{(x-y)^2}{y^2}}{\frac{x-2}{y}} = \frac{(x-2)^2}{y^2} \cdot \frac{y}{x-2}$$

$$= \frac{y(x-2)^2}{y^2(x-2)}$$

$$= \frac{x-2}{y} \qquad y \neq 0, x \neq 2$$

EXAMPLE 3

$$\frac{\frac{x^2-y^2}{xy}}{\frac{(x-y)^2}{2x^2y^2}} = \frac{x^2-y^2}{xy} \cdot \frac{2x^2y^2}{(x-y)^2}$$

$$= \frac{(x+y)(x-y)}{xy} \cdot \frac{2x^2y^2}{(x-y)^2}$$

$$= \frac{(x+y)(x-y)(2x^2y^2)}{xy(x-y)^2}$$

$$= \frac{2xy(x+y)}{x-y} \qquad x \neq y, x \neq 0, y \neq 0$$

EXERCISES 4.5

Divide. Assume that there are no zero divisors.

1. $\dfrac{\frac{x^2y}{z}}{\frac{x^2y^2}{z}}$

2. $\dfrac{\frac{ab}{c^2d^2}}{\frac{2a^2b^2}{c^3}}$

4.6 DIVISION OF RATIONAL EXPRESSIONS

3. $\dfrac{\dfrac{12ab}{5c^3}}{\dfrac{8a^2b^2}{15c^2d}}$

4. $\dfrac{\dfrac{-24x^3y^2}{7b^2}}{\dfrac{-12x^2}{35b^3}}$

5. $\dfrac{\dfrac{36a^3b^3c^3}{x^2y^2}}{\dfrac{24a^2b^2}{5x^2y^3}}$

6. $\dfrac{\dfrac{x+y}{x-y}}{\dfrac{x+y}{x^2-y^2}}$

7. $\dfrac{\dfrac{a^2-ab}{ab}}{\dfrac{3a-3b}{a^2b^2}}$

8. $\dfrac{\dfrac{3a-3}{a}}{\dfrac{a^2-1}{a^2b^2}}$

9. $\dfrac{\dfrac{x^2-9}{x^2-4}}{\dfrac{3x-9}{4x-8}}$

10. $\dfrac{\dfrac{x^2y^2-4x^2}{2z^3}}{\dfrac{2x-xy}{10z^5}}$

11. $\dfrac{\dfrac{25-y^2}{4-y^2}}{\dfrac{25-5y}{12-6y}}$

12. $\dfrac{\dfrac{x^2+x-12}{x^2-1}}{\dfrac{3x+12}{4x^3+4x^2}}$

13. $\dfrac{\dfrac{2y^2-y-28}{3y^2-y-2}}{\dfrac{4y^2+16y+7}{3y^2+11y+6}}$

14. $\dfrac{\dfrac{3x-2x^2}{5x^2+23x-10}}{\dfrac{5x^5-3x^4}{10x^2-9x+2}}$

15. $\dfrac{\dfrac{x^4+64x}{x^3-4x^2}}{\dfrac{x^2-4x+16}{x}}$

16. $\dfrac{\dfrac{y^2+2y-8}{y^2+7y+12}}{\dfrac{y^2-3y+2}{y^2-2y-3}}$

17. $\dfrac{\dfrac{27y^3+8}{6y^2+19y+10}}{\dfrac{9y^2-6y+4}{4y^2-25}}$

18. $\dfrac{\dfrac{5x^3-2x^2}{3x^2-5x+2}}{\dfrac{6-11x-10x^2}{2x^2+x-3}}$

19. $\dfrac{\dfrac{r^2+5r+6}{2r^2+r-1}}{\dfrac{r^2-9}{(r+1)^2}}$

20. $\dfrac{\dfrac{(x^3-27)(x^2-9)}{(x-3)^2(x+3)^2}}{\dfrac{x^2+3x+9}{x^2+3x}}$

4.7 ◆ Complex Rational Expressions

When the division of one rational expression by another is indicated by the fraction bar (as in Section 4.6), that is, when $\dfrac{P}{Q} \div \dfrac{R}{S}$ is indicated by

$$\dfrac{\dfrac{P}{Q}}{\dfrac{R}{S}}$$

where neither Q, R, nor S is the polynomial zero, we call the expression

$$\dfrac{\dfrac{P}{Q}}{\dfrac{R}{S}}$$

a **complex rational expression**. For example,

$$\dfrac{\dfrac{3x}{x+1}}{\dfrac{x^2}{(x+1)^2}}, \quad \dfrac{\dfrac{x-y}{xy}}{\dfrac{x^2-y^2}{x^2y^2}}, \quad \text{and} \quad \dfrac{\dfrac{1}{x}-\dfrac{1}{y}}{\dfrac{1}{x}+\dfrac{1}{y}}$$

are complex rational expressions.

In simplifying complex rational expressions, always remember that the horizontal fraction bar indicates division. Some techniques for simplifying complex rational expressions are illustrated by the examples below.

EXAMPLE 1. Assume that there are no zero divisors.
METHOD 1

$$\dfrac{\dfrac{a}{b}+\dfrac{b}{a}}{ab} = \dfrac{\dfrac{a^2+b^2}{ab}}{ab} = \dfrac{a^2+b^2}{ab} \cdot \dfrac{1}{ab} = \dfrac{a^2+b^2}{a^2b^2}$$

METHOD 2

$$\dfrac{\dfrac{a}{b}+\dfrac{b}{a}}{ab} = \dfrac{\dfrac{a}{b}+\dfrac{b}{a}}{ab} \cdot \dfrac{ab}{ab} = \dfrac{\dfrac{a}{b} \cdot ab + \dfrac{b}{a} \cdot ab}{a^2b^2} = \dfrac{a^2+b^2}{a^2b^2}$$

4.7 COMPLEX RATIONAL EXPRESSIONS

EXAMPLE 2. Assume that there are no zero divisors.
METHOD 1

$$\frac{2 + \dfrac{x}{y}}{\dfrac{x}{y^2}} = \frac{\dfrac{2y + x}{y}}{\dfrac{x}{y^2}} = \frac{2y + x}{y} \cdot \frac{y^2}{x} = \frac{2y^2 + xy}{x}$$

METHOD 2

$$\frac{2 + \dfrac{x}{y}}{\dfrac{x}{y^2}} = \frac{2 + \dfrac{x}{y}}{\dfrac{x}{y^2}} \cdot \frac{y^2}{y^2} = \frac{2y^2 + xy}{x}$$

EXAMPLE 3. Assume that there are no zero divisors.
METHOD 1

$$\frac{\dfrac{1}{x-y} + \dfrac{1}{x+y}}{\dfrac{1}{x-y} - \dfrac{1}{x+y}} = \frac{\dfrac{(1)(x+y)}{(x-y)(x+y)} + \dfrac{(1)(x-y)}{(x+y)(x-y)}}{\dfrac{(1)(x+y)}{(x-y)(x+y)} - \dfrac{(1)(x-y)}{(x+y)(x-y)}}$$

$$= \frac{\dfrac{x+y}{x^2-y^2} + \dfrac{x-y}{x^2-y^2}}{\dfrac{x+y}{x^2-y^2} - \dfrac{x-y}{x^2-y^2}}$$

$$= \frac{\dfrac{x+y+x-y}{x^2-y^2}}{\dfrac{x+y-x+y}{x^2-y^2}}$$

$$= \frac{\dfrac{2x}{x^2-y^2}}{\dfrac{2y}{x^2-y^2}}$$

$$= \frac{2x}{x^2-y^2} \cdot \frac{x^2-y^2}{2y}$$

$$= \frac{2x(x^2-y^2)}{2y(x^2-y^2)}$$

$$= \frac{x}{y}$$

METHOD 2

$$\frac{\dfrac{1}{x-y}+\dfrac{1}{x+y}}{\dfrac{1}{x-y}-\dfrac{1}{x+y}} = \frac{\dfrac{1}{x-y}+\dfrac{1}{x+y}}{\dfrac{1}{x-y}-\dfrac{1}{x+y}} \cdot \frac{(x-y)(x+y)}{(x-y)(x+y)}$$

$$= \frac{(x+y)+(x-y)}{(x+y)-(x-y)} = \frac{2x}{2y} = \frac{x}{y}$$

EXERCISES 4.6

Simplify. Assume that there are no zero divisors.

1. $\dfrac{\dfrac{3ab}{c}}{\dfrac{6ab}{c^2}}$

2. $\dfrac{\dfrac{27x^2y^2}{7}}{\dfrac{9x}{14}}$

3. $\dfrac{\dfrac{x}{y}}{\dfrac{x^2}{y^3}}$

4. $\dfrac{1+\dfrac{x}{y}}{\dfrac{x^2}{y^2}}$

5. $\dfrac{\dfrac{a^2}{b^2}-1}{\dfrac{a-b}{3}}$

6. $\dfrac{\dfrac{x^2}{2}-\dfrac{3xy}{4}}{\dfrac{x}{6}-\dfrac{y}{4}}$

7. $\dfrac{1+\dfrac{b}{a+b}}{1+\dfrac{3b}{a-b}}$

8. $\dfrac{\dfrac{1}{8c}-\dfrac{d}{4c^2}}{\dfrac{1}{4}-\dfrac{d}{2c}}$

9. $\dfrac{2+\dfrac{3y}{x-y}}{2-\dfrac{3y}{x+2y}}$

10. $\dfrac{x+\dfrac{9}{x}-6}{\dfrac{x}{2}-\dfrac{9}{2x}}$

11. $\dfrac{\dfrac{1}{x}+\dfrac{1}{y}}{\dfrac{1}{x}-\dfrac{1}{y}}$

12. $\dfrac{\dfrac{1}{a}+\dfrac{1}{b}+\dfrac{1}{c}}{\dfrac{1}{a}-\dfrac{1}{b}-\dfrac{1}{c}}$

13. $\dfrac{x+4+\dfrac{4}{x}}{x-1-\dfrac{6}{x}}$

14. $\dfrac{\dfrac{x+1}{x-1}+\dfrac{x-1}{x+1}}{\dfrac{x+1}{x-1}-\dfrac{x-1}{x+1}}$

15. $\dfrac{\dfrac{x^2+y^2}{x^2-y^2}}{\dfrac{x-y}{x+y}-\dfrac{x+y}{x-y}}$

16. $\dfrac{\dfrac{a}{a+b}+\dfrac{b}{a-b}}{\dfrac{b}{a+b}-\dfrac{a}{a-b}}$

17. $\dfrac{\dfrac{1}{x+y}+\dfrac{1}{x-y}}{\dfrac{1}{x+y}-\dfrac{1}{x-y}}$

18. $\dfrac{\dfrac{6}{x^2-4}+\dfrac{2}{2-x}}{\dfrac{3}{x^2-4}+\dfrac{2}{x+2}}$

CHAPTER 4 REVIEW

1. Specify any real values of the variables for which the following rational expressions are undefined.

   (a) $\dfrac{2}{x}$ (b) $\dfrac{3}{y-2}$ (c) $\dfrac{x-4}{x-3}$ (d) $\dfrac{x}{x^2-1}$

Simplify (Exercises 2–5).

2. $\dfrac{5x}{35x^2}$

3. $\dfrac{3x^2-3}{x-1}$

4. $\dfrac{x^3-8}{x^2+2x+4}$

5. $\dfrac{x^2-x-6}{x^2-10x+21}$

Add (Exercises 6–9).

6. $\dfrac{x+1}{y}+\dfrac{x-1}{y}$

7. $\dfrac{a-1}{2b}+\dfrac{a}{2b}$

8. $\dfrac{7}{5x-10}+\dfrac{-5}{3x-6}$

9. $\dfrac{x}{x^2-16}+\dfrac{x+1}{x^2-5y+4}$

Subtract (Exercises 10–13).

10. $\dfrac{y+1}{x}-\dfrac{y-1}{x}$

11. $\dfrac{3}{x+2}-\dfrac{x+3}{x-2y}$

12. $\dfrac{2}{x+2}-\dfrac{3}{x+3}$

13. $\dfrac{8}{x^2-4y^2}-\dfrac{2}{x^2+5xy+6y^2}$

Multiply (Exercises 14–15).

14. $\dfrac{3a}{4ab-6b^2}\cdot\dfrac{2a-3b}{12a}$

15. $\dfrac{4a^2+8a+3}{2a^2-5a+3}\cdot\dfrac{6a^2-9a}{4a^2-1}$

Divide (Exercises 16–17).

16. $\dfrac{\dfrac{x^2 - xy}{xy}}{\dfrac{2x - 2y}{xy}}$

17. $\dfrac{\dfrac{a^2 + a - 2}{a^2 + 2a - 3}}{\dfrac{a^2 + 7a + 10}{a^2 - 2a - 15}}$

Simplify (Exercises 18–20).

18. $\dfrac{\dfrac{1}{8a} - \dfrac{d}{4c^2}}{\dfrac{1}{4} - \dfrac{d}{2c}}$

19. $\dfrac{\dfrac{a^2}{2} - \dfrac{3ab}{4}}{\dfrac{a}{6} - \dfrac{b}{4}}$

20. $\dfrac{\dfrac{a - 2b}{2a - b}}{\dfrac{1}{a + b} - \dfrac{1}{2a - b}}$

CHAPTER 5

First-Degree Equations and Inequalities

5.1 ◆ Algebraic Equations

Sentences such as

The sum of eight and two is ten.
The product of five and six is twenty-five.

are called statements. A **statement** is a sentence that may be labeled true or false. The first statement above is true; the second false.

Some sentences have statement form, but are not statements because, as they stand, they cannot be labeled either true or false. For example,

That number is less than five.
He is twenty-one years old.

have statement form but are not statements because they cannot be labeled either true or false. Note that when "that number" and "he" in the given sentences are replaced by a specified name, these sentences become statements. Sentences such as those above are called **open sentences.**

Mathematical sentences, whether statements or open sentences, containing the verb "equals" or the verb phrase "is equal to" are called **equations.** For example,

$$3 + 8 = 11$$
$$5 + 8 = 7$$
$$x - 5 = 9$$
$$2x - 6 = 8$$

are all equations.

A set of numbers that are either implicitly or explicitly stated as permissible replacements of a variable in an open sentence is called the **replace-**

ment set or **domain** of that variable. In this book, unless otherwise stated, the domain will be the set of real numbers. The subset of the replacement set whose elements make an equation a true statement is called the **solution set** of the equation. Every element of the solution set is called a **solution** and is said to **satisfy** the equation. For example, if we replace x by 2 in the equation

$$x + 6 = 8$$

we obtain the true statement $2 + 6 = 8$, and 2 is a solution of this equation. If we replace x by any real number other than 2, we obtain a false statement. Thus the solution set of this equation is $\{2\}$.

5.2 • Equivalent Equations

Two or more algebraic equations that have the same solution set are called **equivalent equations**. Thus

$$3x + 4 = x + 8$$
$$2x = 4$$
$$x = 2$$

are equivalent equations because $\{2\}$ is the solution set of each equation.

A **first degree algebraic equation**, called a **linear** equation, is equivalent to an equation of the form

$$x = k$$

where k is a real number. Examples of linear equations are

$$x + \tfrac{2}{3} = \tfrac{5}{6}$$
$$\tfrac{1}{2}(x - 7) = \tfrac{1}{3}(3x + 5)$$
$$5y - \tfrac{7}{2} = 3y$$

In finding the solution set of a first-degree equation we make extensive use of certain axioms of the equality relation given in Chapter 1, Section 1.2. We restate them here.

E-1. (*Addition Property*) If $a=b$, then $a+c=b+c$ and $c+a=c+b$.
E-2. (*Subtraction Property*) If $a=b$, then $a-c=b-c$ and $c-a=c-b$.
E-3. (*Multiplication Property*) If $a = b$, then $ac = bc$ and $ca = cb$.
E-4. (*Multiplication Cancellation Property or Division Property*) If $ac = bc$ or $ca = cb$, $c \neq 0$, then $a = b$.

We apply these properties to a linear equation to obtain a sequence of equivalent equations until we obtain an equation of the form $x = k$. Study the examples below.

5.3 SOLVING EQUATIONS

EXAMPLE 1. Find an equation of the form $x = k$, k a real number, equivalent to $3x + 4 = x + 8$.

SOLUTION. We use the axioms of equality to find this sequence of equivalent equations:

$3x + 4 = x + 8$	
$2x + 4 = 8$	subtracting x from both members
$2x = 4$	subtracting 4 from both members
$x = 2$	dividing both members by 2

EXAMPLE 2. Find an equation of the form $x = k$, k a real number, equivalent to $\frac{1}{2}y + 32 = 35$.

SOLUTION. $\frac{1}{2}y + 32 = 35$

$\frac{1}{2}y = 3$	subtracting 32 from both members
$y = 6$	multiplying both members by 2

When the axioms of equality are applied as above they are called **elementary transformations**.

5.3 ◆ Solving Equations

The application of the axioms of equality stated in the previous section permits us to transform an equation, whose solution set is not obvious, through a sequence of equivalent equations into an equation that has an obvious solution set. For example, let us consider the equation

$$2x - 3 = x + 4$$

If we add $3 - x$ to both members we obtain the equivalent equation

$$x = 7$$

whose solution set is $\{7\}$.

Care must be exercised in the application of the multiplication and division properties to be sure we have excluded multiplication and division by 0. For example, to find the solution set of the equation

(1) $$\frac{x}{x-4} = \frac{4}{x-4} + 2$$

we might multiply each member by $x - 4$ to find a supposedly equivalent equation which has no fractions. We would then have

$$(x-4)\frac{x}{x-4} = (x-4)\frac{4}{x-4} + 2(x-4)$$

or

(2) $$x = 4 + 2x - 8$$

which *is* equivalent to

(3) $$x = 4$$

It would appear that 4 is a solution of Equation 1. But when we replace x by 4 in Equation 1, we have

$$\tfrac{4}{0} = \tfrac{4}{0} + 2$$

which is not a true statement, since $\tfrac{4}{0}$ is not a real number. In obtaining Equation 2, each member of Equation 1 was multiplied by $x - 4$. If we replace x by 4, $x - 4$ becomes 0 and Equations 1 and 2 are not equivalent because the division property is not applicable. Equation 2 is not equivalent to Equation 1, since the solution set of Equation 2 is $\{4\}$ and the solution set of Equation 1 is $\emptyset$.

In this chapter we shall be concerned with first degree equations. Any equation that, through the use of elementary transformations, is equivalent to an equation of the form

(1) $$ax + b = 0$$

where $a \neq 0$, is called a **first degree equation** in one variable. We now show that this equation has one and only one solution.

If we subtract b from both members of Equation 1, we obtain the equivalent equation

(2) $$ax = -b$$

If we divide both members of Equation 2 by a (which is not 0) we obtain

$$x = -\frac{b}{a}$$

The real number $-\dfrac{b}{a}$ is a solution of the equation $ax + b = 0$, since

$$a\left(-\frac{b}{a}\right) + b = 0$$

is a true statement.

To prove that there is one and only one solution, let us suppose there are two solutions, s_1 and s_2, to the equation $ax + b = 0$. We then have

$$as_1 + b = 0 \quad \text{and} \quad as_2 + b = 0$$

5.3 SOLVING EQUATIONS

From this we see by the transitive property of equality that

$$as_1 + b = as_2 + b$$

Subtracting b from both members of this true equation, we have

$$as_1 = as_2$$

Dividing both members by a, which is not 0, we have

$$s_1 = s_2$$

and the two solutions are, in fact, the same. We therefore conclude that the solution set of the equation, $ax + b = 0$, is $\left\{-\dfrac{b}{a}\right\}$.

EXAMPLE 1. Find the solution set of $6x + 4 = 4x - 2$.

SOLUTION. Subtracting $4 + 4x$ from each member of the given equation, we obtain the equivalent equation

$$2x = -6$$

Dividing each member of the above equation by 2, we obtain the equivalent equation

$$x = -3$$

The solution set is $\{-3\}$.

CHECK

$$6(-3) + 4 = 4(-3) - 2$$
$$-18 + 4 = -12 - 2$$
$$-14 = -14$$

Equations in which the absolute value symbol appears require extra attention, since they are not first degree equations. We recall from the definition of the absolute value of a real number that the equation

$$|x| = 2$$

has solution set $\{-2, 2\}$ because $|2| = 2$ and $|-2| = 2$. Thus the equation $|x| = 2$ is equivalent to the open sentence

$$x = 2 \quad \text{or} \quad -x = 2$$

This observation helps us to solve more complicated equations involving the absolute value symbol.

EXAMPLE 2. Find the solution set of $|x - 2| = 8$.

SOLUTION. This equation is equivalent to the open sentence

$$x - 2 = 8 \quad \text{or} \quad -(x - 2) = 8$$

Transforming:

$$(x - 2) + 2 = 8 + 2 \quad \text{or} \quad (-x + 2) - 2 = 8 - 2$$
$$x = 10 \quad \text{or} \quad -x = 6$$
$$x = 10 \quad \text{or} \quad x = -6$$

The solution set is $\{-6, 10\}$.

EXAMPLE 3. Find the solution set of $|2x + 3| = 5$.

SOLUTION. This equation is equivalent to the open sentence

$$2x + 3 = 5 \quad \text{or} \quad -(2x + 3) = 5$$

Transforming:

$$2x + 3 = 5 \quad \text{or} \quad -2x - 3 = 5$$
$$(2x + 3) - 3 = 5 - 3 \quad \text{or} \quad (-2x - 3) + 3 = 5 + 3$$
$$2x = 2 \quad \text{or} \quad -2x = 8$$
$$x = 1 \quad \text{or} \quad x = -4$$

The solution set is $\{-4, 1\}$.

EXERCISES 5.1

After each equation below is a numeral in parentheses. Test to see whether the number named is a solution of the equation (Exercises 1–8).

1. $3x + 5 = 15$ (9)
2. $5x - 7 = 3$ (2)
3. $2x + 3 = 17$ (7)
4. $4x - 9 = 8$ (3)
5. $x^2 - 4x + 5 = 0$ (5)
6. $x^2 + 2x - 3 = 0$ (1)
7. $x^2 - 2x - 15 = 0$ (−3)
8. $x^3 - 8 = 0$ (−2)

Find the solution sets of the following. Assume no zero divisors (Exercises 9–48).

9. $x + 7 = 10$
10. $x - 3 = 4$
11. $6 = 2 + x$
12. $6x = 18$
13. $x - 2 = 0$
14. $7x - 2 = 4 - x$
15. $6x - 8 = 6 + 4x$
16. $6y + 4 = y + 4$
17. $5x + 7 = 3x + 19$
18. $6 - x = 6x - 8$

5.4 SOLVING EQUATIONS FOR SPECIFIED SYMBOLS

19. $32x - 6 = 4x + 1$
20. $4(3x - 4) - 8 = 0$
21. $2x + 2(x - 2) - 4 = 0$
22. $x^2 - (x - 2)(x - 3) = 19$
23. $x^2 - (x - 2)(x - 3) = 1$
24. $3(x - 2) - 2(x + 3) = -5$
25. $y^2 + 3 - (y + 1)(y + 2) = 0$
26. $\dfrac{x + 6}{5} = 2$
27. $\dfrac{x}{4} + 3 = x - 3$
28. $\dfrac{x - 6}{4} + 3 = x - 12$
29. $\dfrac{x}{2} + \dfrac{x + 3}{4} = -2$
30. $\dfrac{7x - 5}{2} - \dfrac{5x + 4}{3} - 9 = 0$
31. $\dfrac{1}{y} = \dfrac{1}{2}$
32. $\dfrac{3}{x + 4} - \dfrac{2}{x + 2} = 0$
33. $|x| = 7$
34. $|x + 3| = 4$
35. $|4y| = 16$
36. $|x + 1| = 6$
37. $|2x + 5| = 7$
38. $|s + 1| = \tfrac{1}{2}$
39. $|x - 6| = 18$
40. $|\tfrac{1}{4}x + 8| = 16$
41. $\dfrac{2}{2x - 4} = \dfrac{3}{4x}$
42. $\dfrac{3}{2x - 5} - \dfrac{1}{2x - 1} = 0$
43. $\dfrac{3}{x} - \dfrac{1}{2} = \dfrac{6}{x}$
44. $\dfrac{1}{3 - 3x} + \dfrac{1}{2x - 1} = \dfrac{2}{3x - 3}$
45. $7 = -\dfrac{y + 3}{y}$
46. $\dfrac{6}{x - 7} = 4$
47. $\dfrac{5x}{4} - \dfrac{1}{2} = \dfrac{x}{2} + \dfrac{1}{4}$
48. $\dfrac{3}{y - 1} = \dfrac{5}{y + 1}$

5.4 ◆ Solving Equations for Specified Symbols

An equation containing more than one variable or containing symbols such as $a, b, c, \ldots$, which represent constants, can be solved for one of the symbols in terms of the others by using the methods developed in the previous sections of this chapter.

EXAMPLE 1. Solve $a = y - 1$ for y.

SOLUTION. Adding 1 to both members gives

$$a + 1 = y - 1 + 1$$
$$a + 1 = y$$
$$y = a + 1$$

EXAMPLE 2. Solve $I = prt$ for r, where $p \neq 0$ and $t \neq 0$.

SOLUTION. Dividing both members by pt gives

$$r = \frac{I}{pt}$$

EXAMPLE 3. Solve $c = \dfrac{H}{m(T-t)}$ for T, where $c \neq 0$, $m \neq 0$, and $T \neq t$.

SOLUTION

$mc(T-t) = H$	multiplying both members by $m(T-t)$
$T - t = \dfrac{H}{mc}$	dividing both members by mc
$T = \dfrac{H}{mc} + t$	adding t to both members

EXAMPLE 4. Solve $K = \tfrac{1}{2}mv^2 - mh$ for h, where $m \neq 0$.

$K - \tfrac{1}{2}mv^2 = -mh$	subtracting $\tfrac{1}{2}mv^2$ from both members
$h = \dfrac{K - \tfrac{1}{2}mv^2}{-m}$	dividing both members by $-m$

EXERCISES 5.2

Solve each equation for the symbol in parentheses. Assume no zero divisors.

1. $x + b = 1$ (x)
2. $by - 2 = m$ (y)
3. $ax + bx = k$ (x)
4. $2x + a = c$ (x)
5. $\dfrac{1}{a}x = \dfrac{1}{b}$ (x)
6. $c - dx = m + x$ (m)
7. $f = am$ (m)
8. $EM = W$ (M)
9. $v = \tfrac{1}{2}gt^2$ (g)
10. $S = \tfrac{1}{3}\pi r^2 h$ (h)
11. $\dfrac{W_1}{W_2} = \dfrac{d_2}{d_1}$ (W_1)
12. $M = \dfrac{W}{E}$ (E)
13. $d = \dfrac{l^2 W}{8s}$ (W)
14. $2s = a + b + c$ (a)
15. $E = IR$ (R)
16. $W = \dfrac{2\pi lRP}{dr}$ (R)
17. $F = \tfrac{9}{5}C + 32$ (C)
18. $l = ar^{n-1}$ (a)
19. $S = at - \tfrac{1}{2}gt^2$ (g)
20. $A = \dfrac{h}{2}(b + b')$ (b)
21. $S = \dfrac{a + ar^n}{1 - r}$ (a)
22. $S = \dfrac{a}{1 - r}$ (r)
23. $d = \dfrac{1 - a}{n - 1}$ (n)
24. $I = \dfrac{ne}{R + nr}$ (n)
25. $\dfrac{1}{R} = \dfrac{1}{r_1} + \dfrac{1}{r_2}$ (r_1)

5.5 ◆ Identities

There are two kinds of open sentence equations: **identical equations** and **conditional equations**. Conditional equations are those which are satisfied only by certain elements in the replacement sets of the variables. The equations we have studied in the previous sections of this chapter are conditional equations.

An identical equation, or **identity,** is one that is satisfied for every element in the replacement sets of the variables. The following are examples of identities:

$$x^2 - 1 = (x + 1)(x - 1)$$
$$x^2 + 2x + 1 = (x + 1)^2$$
$$x^2 - 4x + 3 = (x - 1)(x - 3)$$

We determine whether a given equation is or is not an identity by transforming one or both members of the equation to obtain a sequence of equivalent equations which ends with an obvious identity (that is, an equation whose two members are identical) or an equation that is obviously conditional. For example,

$$(x + 4)(x - 4) = (x + 3)(x - 3) - 7$$
$$(x + 4)(x - 4) = x^3 - 9 - 7$$
$$(x + 4)(x - 4) = x^2 - 16$$
$$(x + 4)(x - 4) = (x + 4)(x - 4)$$

Thus we find that the original equation is an identity.

Now consider the equation

$$(x - 3)^2 = x^2 + 6x + 9$$

After multiplying we get

$$x^2 - 6x + 9 = x^2 + 6x + 9$$

from which we get the equivalent equations

$$12x = 0$$
$$x = 0$$

The equation $x = 0$ is a conditional equation since only 0 satisfies it. Hence the given equation is a conditional equation since it is equivalent to $x = 0$.

In verifying that an equation is an identity it is most desirable to transform one member of the equation to the other member which is left unaltered. The member we transform is usually the more complicated one.

FIRST-DEGREE EQUATIONS AND INEQUALITIES

EXAMPLE 1. Verify that $5 - \dfrac{4x - 9}{x - 3} = \dfrac{x - 6}{x - 3}$, $x \neq 3$, is an identity.

SOLUTION. We shall transform the left member of the equation.

$$5 - \dfrac{4x - 9}{x - 3} = \dfrac{5(x - 3)}{x - 3} - \dfrac{4x - 9}{x - 3}$$

$$= \dfrac{5x - 15 - 4x + 9}{x - 3}$$

$$= \dfrac{x - 6}{x - 3}$$

EXAMPLE 2. Verify that

$$x^3 - x(x - 1)^2 + (x - 1)^2 - 3x^2 + 3x = 1$$

is an identity.

SOLUTION. We shall transform the left member of the equation.

$x^3 - x(x - 1)^2 + (x - 1)^2 - 3x^2 + 3x$
$= x^3 - x(x^2 - 2x + 1) + (x^2 - 2x + 1) - 3x^2 + 3x$
$= x^3 - x^3 + 2x^2 - x + x^2 - 2x + 1 - 3x^2 + 3x$
$= 1$

EXERCISES 5.3

Verify that the following equations are identities. Assume there are no zero divisors.

1. $3(x + 2) = 3x + 6$
2. $\dfrac{x}{2} + 3 = \dfrac{x + 6}{2}$
3. $(x + 3)(x - 3) + 9 = x^2$
4. $\dfrac{8x - 6x}{2} + 3 = x + 3$
5. $(x + 2)^2 - 4x = x^2 + 4$
6. $\dfrac{x^2 - 6x + 8}{x - 2} = x - 4$
7. $x(x + 3) = (x + 2)(x + 1) - 2$
8. $\left(x + \dfrac{1}{x}\right)^2 - \left(x - \dfrac{1}{x}\right)^2 = 4$
9. $(3x - 5)^2 - (3x + 5)^2 + 60x = 0$
10. $\dfrac{1}{x + 2} - \dfrac{1}{x + 3} = \dfrac{1}{x^2 + 5x + 6}$
11. $\dfrac{3x}{4} = \dfrac{4 + 5x}{4} - \dfrac{x + 2}{2}$
12. $\dfrac{x^2}{4} + x = \dfrac{(x + 2)^2}{4} - 1$
13. $\dfrac{x^2 + x - 6}{9 - x^2} = \dfrac{x - 2}{3 - x}$
14. $\dfrac{a^2 + 3a - 4}{a^2 - 4a + 3} = \dfrac{a + 4}{a - 3}$
15. $\dfrac{1}{x} + \dfrac{2}{x - 2} = \dfrac{3x - 2}{x^2 - 2x}$
16. $\dfrac{x}{x^2 - 1} - \dfrac{3}{x - 1} = \dfrac{2x + 3}{1 - x^2}$

17. $a + b - \dfrac{2ab}{a+b} = \dfrac{a^2 + b^2}{a+b}$

18. $\dfrac{x^3 - y^3}{x - y} = x^2 + xy + y^2$

19. $\dfrac{2}{3}(x - 4) + \dfrac{1}{3}(x - 3) = x - \dfrac{11}{3}$

20. $\dfrac{a}{ab - b^2} - \dfrac{b}{a^2 - ab} = \dfrac{a+b}{ab}$

5.6 ◆ Inequalities

Open sentences of the form

$$3 + x < 9, \quad x + 1 \leq 5, \quad 2x - 3 > 4, \quad 5x - 2 \geq 3$$

are called **inequalities**. For particular values of the variable, both members are real numbers and the inequality is a statement that one real number is less than ($<$), less than or equal to ($\leq$), greater than ($>$), or greater than or equal to ($\geq$) another real number.

Any element of the replacement set of the variable for which the inequality is a true statement is called a **solution** of the inequality, and the set of all solutions is called the **solution set** of the inequality.

In order to find solution sets of inequalities, we shall need some fundamental properties of inequalities. We shall accept these properties without proof.

I-1. *(Transitive Property)* If $a < b$ and $b < c$, then $a < c$.

I-2. *(Addition Property)* If $a < b$, then $a + c < b + c$.

I-3. *(Multiplication Property for Positive Numbers)* If $a < b$ and $c > 0$, then $ac < bc$.

I-4. *(Multiplication Property for Negative Numbers)* If $a < b$ and $c < 0$, then $ac > bc$.

EXAMPLE 1. Find the solution set of $x + 6 < 9$.

SOLUTION. Adding -6 to both members of the inequality, we have the equivalent inequality $x < 3$. This means that any real number less than 3 is a solution of the given inequality, so that the solution set is the set of all real numbers less than 3.

We can graph this solution set on the number line as shown in Figure 5.1. The heavy trace is the **graph** of the solution set. Notice the open circle around the point with label 3. This tells us that 3 is

Figure 5.1

FIRST-DEGREE EQUATIONS AND INEQUALITIES

not included in the solution set. The arrow at the end of the heavy trace tells us that all numbers which label points to the left of the heavy trace are also solutions.

We can use set-builder notation to designate solutions of inequalities. The solution set of the above inequality may be written

$$\{x \mid x < 3\}$$

This symbol is read "The set of all x such that x is less than 3."

EXAMPLE 2. Find the solution set of $-6 < 2x < 12$.

SOLUTION. This inequality is read as "$2x$ is less than 12 and greater than -6" or "-6 is less than $2x$, which is less than 12." We find the solution set of this inequality by multiplying each member by $\frac{1}{2}$ and getting the equivalent inequality

$$-3 < x < 6$$

The solution set is

$$\{x \mid -3 < x < 6\}$$

The graph of this solution set is shown in Figure 5.2.

Figure 5.2

EXAMPLE 3. Find the solution set of $-2 \leq 3x + 1 \leq 10$.

SOLUTION. This inequality is read "$3x + 1$ is greater than or equal to -2 and less than or equal to 10" or "-2 is less than or equal to $3x + 1$, which is less than or equal to 10." Using the addition property for inequality, we add -1 to each member and get the equivalent inequality

$$-3 \leq 3x \leq 9$$

Using the multiplication property for positive numbers, we multiply each member by the positive number $\frac{1}{3}$, giving the equivalent inequality

$$-1 \leq x \leq 3$$

whose solution set is $\{x \mid -1 \leq x \leq 3\}$. The graph of this solution set is shown in Figure 5.3. The filled-in circles at the points labeled

Figure 5.3

−1 and 3 indicate that these numbers are included in the solution set.

EXAMPLE 4. Find the solution set of $-1 < -2x + 3$.

SOLUTION. Using the addition property for inequality, we add -3 to each member and get the equivalent inequality

$$-4 < -2x$$

Using the multiplication property for negative numbers, we multiply both members by the negative number $-\frac{1}{2}$ and get the equivalent inequality

$$2 > x$$

whose solution set is $\{x \mid x < 2\}$.

EXAMPLE 5. Find the solution set of $|x + 1| > 4$.

SOLUTION. We can use the definition of absolute value to get an equivalent open sentence:

$(x + 1)$ is more than 4 units from 0

If we picture this situation on the number line, we see that the point whose coordinate is $x + 1$ is either to the right of the point whose coordinate is 4, that is,

$$x + 1 > 4$$

or the point whose coordinate is $x + 1$ is to the left of the point whose coordinate is -4, that is,

$$x + 1 < -4$$

Thus we have

$$x + 1 > 4 \quad \text{or} \quad x + 1 < -4$$

We now use the addition property of inequality in adding -1 to the members of both inequalities to get the equivalent open sentence

$$x > 3 \quad \text{or} \quad x < -5$$

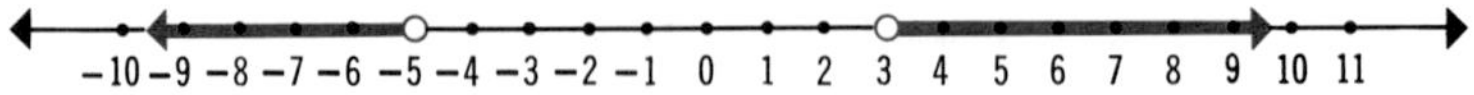

FIRST-DEGREE EQUATIONS AND INEQUALITIES

Figure 5.4

We conclude that the solution set of $|x + 1| > 4$ is

$$\{x \mid x > 3 \quad \text{or} \quad x < -5\}$$

The graph of this solution set is shown in Figure 5.4.

EXAMPLE 6. Find the solution set of $|x + 1| < 4$.

SOLUTION. Again we use the definition of absolute value to get the equivalent open sentence:

$$x + 1 \text{ is within 4 units of 0}$$

If we picture this situation on the number line we see that the point whose coordinate is $x + 1$ is between the points whose coordinates are 4 and -4. Thus

$$x + 1 < 4 \quad \text{and} \quad x + 1 > -4$$

We write this

$$-4 < x + 1 < 4$$

Using the addition property of inequalities and adding -1 to all members of the above inequality, we get the equivalent inequality

$$-5 < x < 3$$

whose solution set is $\{x \mid -5 < x < 3\}$. The graph of this solution set is shown in Figure 5.5.

Figure 5.5

EXERCISES 5.4

Find the solution sets. Graph each solution set.

1. $x > -3$
2. $4x < 12$
3. $x + 6 > -3$
4. $x - 3 > 6$

5. $3 + x \leq 5$
6. $2x + 1 > 1$
7. $2x - 5 > 3$
8. $5x - 2 > 8$
9. $2 - x < 4$
10. $24 < x - 18$
11. $6 \leq x \leq 12$
12. $0 \leq 1 - x \leq 1$
13. $\dfrac{2x - 5}{3} \leq 3$
14. $\dfrac{x - 1}{2} < 3$
15. $\dfrac{8 - 2x}{4} > 4$
16. $-5 \leq 2x + 1 \leq 4$
17. $\dfrac{x + 4}{5} \leq 1$
18. $|x + 1| > 4$
19. $|x - 1| > 4$
20. $2|x + 1| < 6$
21. $|2x + 1| \leq 7$
22. $2|2x + 2| \leq 8$
23. $3|x - 4| \geq 9$
24. $\left|\dfrac{x - 1}{3}\right| > 2$
25. $\left|\dfrac{1 - 2x}{5}\right| \leq 3$

5.7 ♦ Applications of Equations and Inequalities

We shall now illustrate the use of equations and inequalities in solving "word problems." To solve such problems, we first change the verbal statements into symbolic mathematical sentences. That is, we construct a mathematical model. This is the most difficult part of solving word problems, and there is no single means available to do this. The following suggestions should, however, prove helpful.

1. Read the problem carefully until you understand it.
2. Determine the quantities asked for and represent them by symbols.
3. Write the conditions of the problem in the form of a mathematical sentence, using the proper mathematical symbol for the verb used in the English sentence.

EXAMPLE 1. Find three consecutive integers whose sum is 84.

SOLUTION. Let x represent the smallest of the three integers. Then

$x + 1$ represents the second integer

and

$x + 2$ represents the third integer

The conditions of the problem may be written

$$x + (x + 1) + (x + 2) = 84$$

This equation is equivalent to

$$3x + 3 = 84$$
$$3x = 81$$
$$x = 27$$

The three integers are 27, 28, and 29.

CHECK

$$27 + 28 + 29 = 84$$

EXAMPLE 2. A jar is filled with pennies and nickels. It contains three times as many nickels as pennies. The total amount of money in the jar is $19.20. How many coins of each kind are in the jar?

SOLUTION. Let x represent the number of pennies in the jar. Then $3x$ will represent the number of nickels. The value of each penny is 1 cent so that the value of the pennies in the jar is x cents. The value of each nickel is 5 cents so that the value of the nickels in the jar is $5(3x)$ cents. The total amount of money in the jar is 1920 cents. The conditions of the problem may then be written as

$$x + 5(3x) = 1920$$

This equation is equivalent to

$$x + 15x = 1920$$
$$16x = 1920$$
$$x = 120$$

Hence $3x = 360$, and there are 120 pennies and 360 nickels in the jar.

CHECK

$$120 + (5)(3)(120) = 1920$$

EXAMPLE 3. The inlet pipe can fill a pool in 30 minutes, and the outlet pipe can empty it in 20 minutes. If the pool is full and both pipes are opened by mistake, how long will it take before the pool will be empty?

SOLUTION. Let x represent the part emptied in one minute with both pipes open. Then $\frac{1}{30}$ is the part filled in one minute by the inlet pipe, $\frac{1}{20}$

5.7 APPLICATIONS OF EQUATIONS AND INEQUALITIES

is the part emptied in one minute by the outlet pipe, and $\frac{1}{x}$ is the part emptied in one minute with both pipes opened. The conditions of the problem can be stated as

$$\frac{1}{x} = \frac{1}{20} - \frac{1}{30}$$

This equation is equivalent to

$$60 = 3x - 2x$$
$$x = 60$$

Hence it will take 60 minutes before the pool is empty.

CHECK

$$\tfrac{1}{60} = \tfrac{1}{20} - \tfrac{1}{30}$$

EXAMPLE 4. A gourmet food store grinds its own coffee. The owner mixes coffee worth 70¢ a pound with another blend worth 80¢ a pound to make a mixture worth 78¢ a pound. How many pounds of each blend of coffee are mixed to make 100 pounds of the mixture?

SOLUTION. Let x represent the number of pounds of the 70¢ coffee used. Then $100 - x$ represents the number of pounds of the 80¢ coffee used. The value of x pounds of 70¢ coffee is $70x$ cents. The value of $100 - x$ pounds of 80¢ coffee is $80(100 - x)$ cents. The value of 100 pounds of the mixture, which sells for 78¢ a pound, is $78(100)$ cents. The conditions of the problem can be written as

$$70x + 80(100 - x) = 78(100)$$
$$70x + 8000 - 80x = 7800$$
$$-10x = -200$$
$$x = 20$$
$$100 - x = 80$$

Thus there are 20 pounds of the 70¢ coffee and 80 pounds of the 80¢ coffee mixed together to make 100 pounds of 78¢ coffee.

CHECK

$$20(70) + 80(80) = 1400 + 6400 = 7800$$

EXAMPLE 5. On a trip to a neighboring city, Mr. Slaugh traveled at an average speed of 45 miles per hour. Returning home, his average speed was 55 miles per hour. It took him 6 hours for the round trip. How far was the neighboring city from Mr. Slaugh's home?

SOLUTION. We need to recall that

$$\text{distance} = (\text{rate})(\text{time elapsed})$$

Let t represent the time, in hours, it took Mr. Slaugh to get to the neighboring city. Then $6 - t$ represents the time, in hours, it took for the return trip. Since his outward-bound trip was made at an average speed of 45 miles per hour, the distance was $45t$ miles. The return trip was made at an average speed of 55 miles per hour, so that the distance traveled was $55(6 - t)$ miles. But the two distances were the same, so the conditions of the problem can be stated as

$$45t = 55(6 - t)$$

This is equivalent to

$$45t = 330 - 55t$$
$$100t = 330$$
$$t = 3.3$$

Since it took Mr. Slaugh 3.3 hours to drive to the neighboring city, its distance from his home is $(3.3)45 = 148.5$ miles.

CHECK

$$45(3.3) = 148.5 = 55(2.7)$$

EXAMPLE 6. Maggie's father is twice as old as Maggie. In four years the sum of their ages will be less than 74. What is the oldest Maggie can be?

SOLUTION. Let x represent Maggie's present age, in years. Then $2x$ represents the present age of her father. Four years from now, Maggie's age will be $x + 4$ and her father's age will be $2x + 4$. The conditions of the problem are

$$(x + 4) + (2x + 4) < 74$$
$$3x + 8 < 74$$
$$3x < 66$$
$$x < 22$$

Thus Maggie must be less than 22 years old—that is, the oldest she can be is 21 years old.

5.7 APPLICATIONS OF EQUATIONS AND INEQUALITIES

EXERCISES 5.5

1. Find three consecutive odd integers whose sum is 123. (*Hint.* Odd integers differ by 2.)
2. The sum of three consecutive integers is 93. What is the largest?
3. The larger of two numbers exceeds the smaller by 24. Seven times the smaller is equal to three times the larger. Find the two numbers.
4. One half of some number exceeds one third of that number by 8. Find the number.
5. There are 36 students in a mathematics class. There are twice as many men as women. How many women are in the class?
6. The length of a rectangle is three times its width. The perimeter is 112 inches. Find the dimensions of the rectangle.
7. The difference of the squares of two consecutive integers is 25. Find the integers.
8. The denominator of a fraction is one less than three times the numerator. If 3 is added to both numerator and denominator, the result is $\frac{8}{17}$. Find the original fraction.
9. When the side of a square is increased by 4 inches, the area is increased by 144 square inches. Find the length of the side of the original square.
10. The difference between two numbers is 4. Four times the smaller plus 12 is 6 more than 3 times the larger. Find the numbers.
11. There are nickels, dimes, and quarters in a box. There are twice as many dimes as quarters and $2\frac{1}{2}$ times as many nickels as dimes. The box contains $11.20. How many of each kind of coin are in the box?
12. The admission to a football game was $2.50 for adults and $1 for students. The receipts were $330 for 150 paid admissions. How many adult and students' tickets were sold?
13. A man invested $900, part at 6% and part at $6\frac{1}{2}$%. The income from each investment was the same. How much did he invest at each rate of interest?
14. A man received $23,715 from the sale of a house after the agent had deducted 7% of the sales price as commission. What was the selling price of the house?
15. The Blue Torch Club bought peanuts to sell. The peanuts cost 25¢ for 3 bags. The members ate 10 bags and sold the rest for 10¢ a bag. They made a profit of $4. How many bags of peanuts did the club buy?
16. A collection of nickels and dimes amounts to $6.20. There are 89 coins in all. Find the number of each kind of coin in the collection.
17. A seed dealer mixed grass seed selling at $1.50 a pound with clover seed selling at $1.75 a pound. He wanted a mixture of 100 pounds to sell for $1.60 a pound. How many pounds of each did he use in the mixture?
18. A 75% alcohol solution is 75% alcohol and 25% distilled water. If you have

8 ounces of a solution which is 75% alcohol, how much distilled water must be added to make a mixture that is 30% alcohol?

19. How many gallons of milk containing 4% butterfat must be added to 70 gallons of cream, containing 40% butterfat, to make a mixture containing 18% butterfat?

20. A candy manufacturer has cremes that sell for 80¢ a pound and caramels that sell for $1.60 a pound. How many pounds of each should he use to make an assortment of 200 pounds that sells for $1 a pound?

21. A man can do a piece of work in 10 days, and his son can do it in 15 days. How long will it take them to do the job if they work together?

22. Two stone masons can build a retaining wall in 15 days. If one can do the job alone in 25 days, how long will it take the other to build the retaining wall alone?

23. A tank has two inlet pipes. One can fill the tank in 30 minutes, and the other can fill it in 45 minutes. How long will it take to fill the tank if both pipes are open?

24. Mrs. Peterson takes 4 hours to wash the windows, and her son takes 6 hours to wash them. Working together at their own rates, how long will it take both of them to wash the windows?

25. A tractor can plow a field in 10 days. A team of horses can plow it in 18 days. How long will it take the tractor and the team together to plow the field?

26. Leo can paint his house in 40 hours if he works alone. Deane can do the job in 60 hours. How long will it take them to paint the house if they work together?

27. George left at 9:00 A.M. and walked at the rate of 4 miles per hour. Clarence walked at the rate of 5 miles per hour. At what time will Clarence overtake George if he left from the same place at 9:30 A.M.?

28. Two cars are traveling in the same direction. The average speed of the first car is 50 miles per hour, and the average speed of the second is 65 miles per hour. If the car traveling at 65 miles per hour leaves the same place one hour after the car traveling 50 miles per hour, how long will it take before the second car overtakes the first?

29. Two cities, A and B, are 648 miles apart. One car leaves A traveling toward B. Another car leaves B traveling toward A. The car traveling toward A goes 8 miles per hour faster than the car traveling toward B. The two cars meet at the end of six hours. What is the rate of each?

30. A man weighs 30 pounds more than his teenage son. Their combined weights are less than 351 pounds. What is the maximum that the boy can weigh? (Find the boy's weight to the nearest pound).

31. Mr. Smith and Mr. Stevens agree to furnish at least $25,000 capital to start a new business. If Mr. Smith furnishes $1000 more than Mr. Stevens, what is the least Mr. Stevens can contribute?

32. A student must have an average of 70 to 85% on four tests in order to re-

ceive a grade of C. His grades on the first three tests were 84%, 76%, and 82%. What is the lowest grade on the fourth test that will give him a C?
33. A student must have an average from 95 to 100% on four tests in a course to receive an A. His grades on the first three tests were 98%, 97%, and 90%. What is the lowest grade he could receive on the fourth test and still get an A in the course?
34. Mr. and Mrs. Vines want to build a house with floor area between 1800 and 2500 square feet. The type of house they want costs $15 per square foot to build. Compute the price range of their new home.
35. Dr. Schmidt's mathematics class contains 20 more students than Dr. Larkin's class. If 5 students withdraw from Dr. Schmidt's class and $\frac{1}{3}$ of the students withdraw from Dr. Larkin's class, there are fewer than 86 students left in the two classes. What is the maximum number of students in Dr. Larkin's class before the withdrawal?
36. The Johnsons want to spend between $30,000 and $45,000 to build a house. It costs $15 per square foot to build the type of house they want. What is the largest house they can build?
37. Steve and Henry collect stamps. Steve has three times as many stamps from Iceland as Henry. Together they have fewer than 4000 stamps from Iceland. What is the largest number of Icelandic stamps that Henry can have?

CHAPTER 5 REVIEW

Find the solution sets (Exercises 1–4).

1. $3x + 1 = 7x - 11$
2. $-2(4x - 5) = 4$
3. $\dfrac{x-1}{4} + \dfrac{x-2}{5} = 7$
4. $2x - 6 = 4$

5. Solve $V = \frac{1}{3}\pi r^2 h$ for h
6. Solve $s = v_0 t - \frac{1}{2}gt^2$ for v_0

Find the solutions sets and graph them (Exercises 7–15).

7. $2x + 3 < 5$
8. $-3x + 2 > 11$
9. $-1 \leq 2x + 5 \leq 1$
10. $|x - 3| < 1$
11. $|x - 3| > 1$
12. $|2x - 4| < 6$
13. $|2x - 4| > 6$
14. $|3x - 6| \leq 3$
15. $|3x - 6| \geq 3$

16. The readings on a Fahrenheit thermometer, F, and on a Centigrade thermometer, C, are related by the equations $F = \frac{9}{5}C + 32$. At what temperature do the two thermometers have the same reading?
17. The sum of three consecutive even positive integers is 66. What are the integers?

18. Norbert can complete a job in five hours. David can do the same job in three hours. How long will it take them to do the job if they work together?
19. Ruth has 8 quarts of a solution that is 50% alcohol. How much water does she need to add to get a solution that is 40% alcohol?
20. Wayne collects counties. On his last trip he doubled the number he started with on the outbound portion and added 15 more on his return. He now has more than 25 counties, but fewer than 235. What number of counties could he have had when he started that trip?

CHAPTER **6**

Relations and Functions

6.1 ◆ Cartesian Products

An **ordered pair** of objects, not necessarily different, is a pair of objects considered in a particular order. If we consider a pair of objects a and b, in that order, we designate this ordered pair by the symbol (a, b). We call a, in the ordered pair (a, b), the **first component** of the ordered pair and b the **second component**.

The ordered pair (a, b) is quite different from the set $\{a, b\}$ which contains the two elements a and b. The order in which we list the elements of a set is immaterial. There is no first element of this set. Thus $\{a, b\}$ and $\{b, a\}$ are the same set. In an ordered pair the order in which we list the components is important; thus (a, b) and (b, a) are not names for the same ordered pair.

Two ordered pairs (a, b) and (c, d) are said to be **equal** if and only if $a = c$ and $b = d$. Thus

$$(2, 4) = (1 + 1, 2 + 2)$$
$$(7, \sqrt{9}) = (7, 3)$$

If we restrict ourselves in forming ordered pairs, to choosing components from the set

$$A = \{1, 2, 3\}$$

then it is possible to form nine distinct ordered pairs. For example, the ordered pairs $(2, 1)$, $(2, 2)$, and $(2, 3)$ are all of the ordered pairs possible that have 2 as their first component, and the ordered pairs $(1, 2)$, $(2, 2)$, and $(3, 2)$ are all of the ordered pairs possible that have 2 as their second component. Thus there are five distinct ordered pairs that have 2 as a component.

We can perform all the pairings of the elements of the A in a systematic manner by using a **tree diagram** (Figure 6.1):

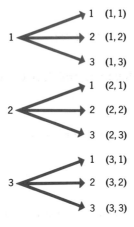

Figure 6.1

Thus we see that it is possible to form nine ordered pairs using the elements of the set A as components. This set of ordered pairs is called the **Cartesian product** of A and A and is designated $A \times A$—reads as "A cross A." Notice that the elements of $A \times A$ are ordered pairs:

$$A \times A = \{(1, 1), (1, 2), (1, 3), (2, 1), (2, 2), (2, 3), (3, 1), (3, 2), (3, 3)\}$$

Now let us form ordered pairs, (a, b), by specifying separately two sets, A and B, one of which is to provide the first components and the other to provide the second components. If

$$A = \{0, 1\} \quad \text{and} \quad B = \{2, 3\}$$

the Cartesian product, $A \times B$, consists of all the ordered pairs whose first components are elements of A and whose second components are elements of B. Using the tree diagram in Figure 6.2, we see that

$$A \times B = \{(0, 2), (0, 3), (1, 2), (1, 3)\}$$

If, on the other hand, we require that all of the first components come from the set B and all of the second components come from the set A, we can form the Cartesian product $B \times A$. The reader should construct his own tree diagram to see that

$$B \times A = \{(2, 0), (2, 1), (3, 0), (3, 1)\}$$

6.1 CARTESIAN PRODUCTS

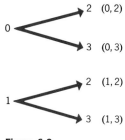

Figure 6.2

This example shows that $A \times B$ and $B \times A$ are not necessarily equal, therefore this operation on sets is *not commutative*. If A is a finite set containing m elements and B is a finite set containing n elements, then $A \times B$ and $B \times A$ are finite sets, each containing mn elements. If A is a finite set containing p elements, then $A \times A$ is a finite set containing p^2 elements.

In the preceding chapter we drew graphs of sets of real numbers by using the number line (Figure 6.3). The number line is a straight line, of indefinite extent, whose points all have real numbers as labels. For each point on this line there is one and only one real number which is its label, and each real number is the label of one and only one point on this line.

Figure 6.3

We now want to be able to draw the graph of sets of ordered pairs of real numbers. To do this we construct the **Cartesian plane** (Figure 6.4). Here we use two number lines, one placed horizontally and the other vertically, which are perpendicular to each other and intersect at the point of each line that has label 0. These number lines are called the **axes** of the Cartesian plane. The horizontal axis is called the **axis of abscissas,** or the **x-axis.** The vertical axis is called the **axis of ordinates,** or the **y-axis.** We can now label every point of this plane with an ordered pair of real numbers. To do this we draw horizontal and vertical lines through each point. We then examine the labels of the points at which these horizontal and vertical lines intersect the two axes (Figure 6.5).

Here, the vertical line through P intersects the horizontal axis at the point labeled -3. We use -3 as the first component of the ordered pair that labels P (Figure 6.6).

The horizontal line through P intersects the vertical axis at the point labeled

RELATIONS AND FUNCTIONS

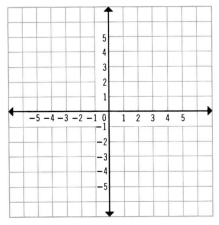

Figure 6.4

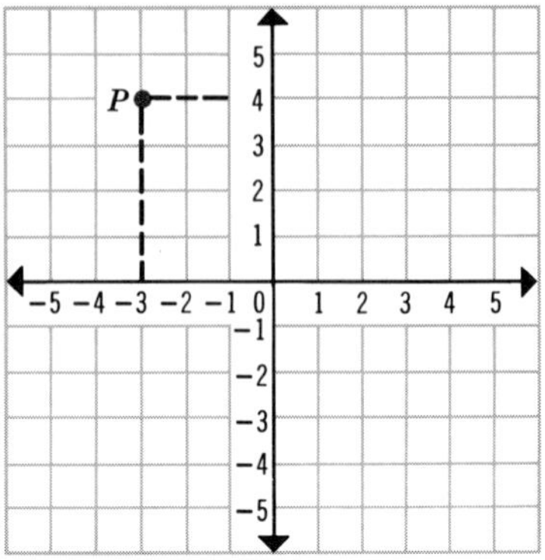

Figure 6.5

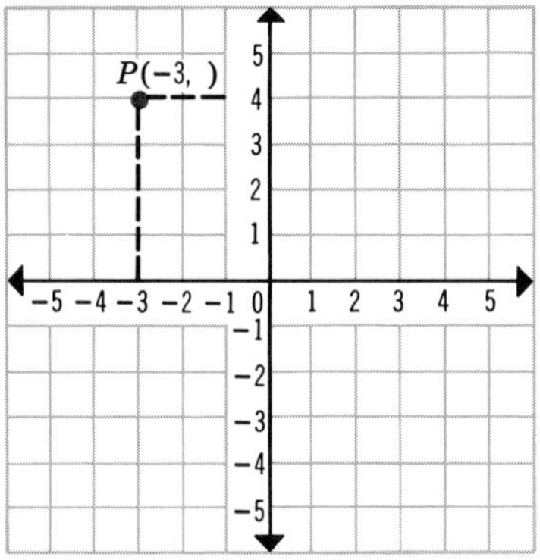

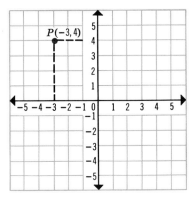

Figure 6.6

Figure 6.7

4. We use 4 as the second component of the ordered pair that labels P (Figure 6.7).

We use the label of the point of intersection of the vertical line through P with the horizontal axis (x-axis) as the first component of the ordered pair which is to label P, and the label of the point of intersection of the horizontal line through P with the vertical axis (y-axis) as the second component.

This manner of labeling points in the Cartesian plane with ordered pairs of real numbers is an example of a **coordinate system.** The points of the axes themselves then have two different labels: the real numbers which were their labels on the number lines and the ordered pairs of real numbers which are their labels as points of the plane. All of the points of the horizontal axis have labels with second component 0. All of the points of the vertical axis have labels with first component 0.

For the set
$$A = \{1, 2, 3\}$$

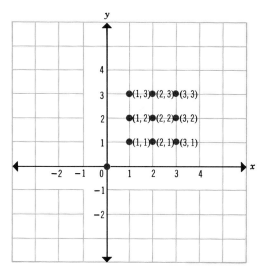

Figure 6.8

the graph of $A \times A$ will be the set of those points of the Cartesian plane whose labels are the ordered pairs of $A \times A$ (Figure 6.8).

EXAMPLE 1. Graph $I \times I$ if $I = \{\ldots, -1, 0, 1, \ldots\}$.

SOLUTION

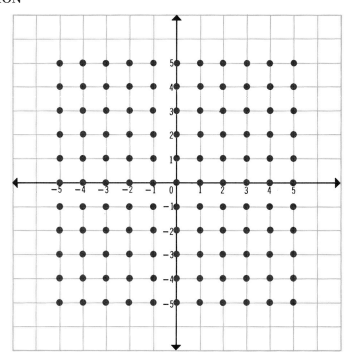

It is impossible to graph all of $I \times I$ because the graph extends indefinitely in all directions. The incomplete graph above indicates the appearance of the graph in a portion of the plane.

EXAMPLE 2. $A = \{x \mid 1 \leq x \leq 2\}$ and $B = \{x \mid 2 \leq x \leq 4\}$. Graph $A \times B$.

SOLUTION. The graph consists of all of the points on the rectangle and inside of the rectangle shown in the figure.

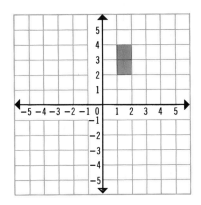

EXAMPLE 3. $P = \{x \mid x \text{ is a real number and } 2 \leq x \leq 3\}$. Graph $P \times P$.

SOLUTION. The graph consists of all of the points on the square and inside of the square shown in the figure.

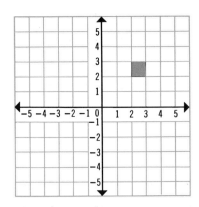

EXAMPLE 4. Graph $R \times R$, where R is the set of all real numbers.

SOLUTION

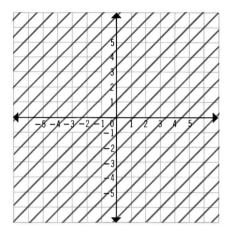

It is impossible to graph all of $R \times R$ because the graph extends indefinitely in all directions. A portion of the graph is shown by shading in the figure. The graph of $R \times R$ is the set of all points of the plane and is called the **real plane**.

EXERCISES 6.1

1. If $A = \{-1, 0, 1\}$, list all the members of $A \times A$.
2. If $A = \{1, 2, 3\}$ and $B = \{-2, -1, 0\}$, list the members of $A \times B$.
3. If $A = \{-1, -2, -3\}$, graph $A \times A$.
4. If $A = \{-1, 0, 1\}$ and $B = \{1, 2, 3\}$, graph $A \times B$.
5. Give the number of members in the Cartesian product set, $A \times A$, if:
   (a) $A = \{a_1\}$ (c) $A = \{a_1, a_2, a_3, a_4\}$
   (b) $A = \{a_1, a_2\}$ (d) $A = \{a_1, a_2, \ldots, a_n\}$

Which of the following statements are true (Exercises 6–10)?

6. If I is the set of integers, then $I \times I$ is the set of all ordered pairs of integers.
7. If R is the set of real numbers, then $R \times R$ is the set of all ordered pairs of real numbers.
8. If $A = \{1, 2, 3\}$ and $B = \{4, 5, 6, 7\}$, then there are seven ordered pairs in $A \times B$.
9. Every element of $A \times A$ is an element of A.
10. $I \times I$ is a subset of $R \times R$.

6.2 ◆ The Cartesian Coordinate System

In the preceding section we saw how each point in the plane received an ordered pair of real numbers as a label. If the ordered pair (x, y) is the label of the point P, the first component, x, of this ordered pair is called the **x-coordinate** or **abscissa** of P and the second component, y, of this ordered pair is called the **y-coordinate** or **ordinate** of P. The two perpendicular number lines are referred to as the **x-axis** (the horizontal number line) and the **y-axis** (the vertical number line). Their point of intersection, which has label $(0, 0)$, is called the **origin**. The entire plane is often referred to as the **x-y plane** and is the graph of the set of all ordered pairs (x, y) that form $R \times R$, the Cartesian product of the set of real numbers by itself. This is referred to as a **Cartesian** or **rectangular coordinate system**.

Every point P in the plane is now associated with an ordered pair, (x, y), of real numbers. The components of the ordered pair are called the **coordinates** of P. The first component is the **x-coordinate** or **abscissa** and the second component is the **y-coordinate** or the **ordinate**.

The two axes of the coordinate system separate the plane into four disjoint sets, called the four **quadrants** of the plane. It is customary to name these quadrants as shown in Figure 6.9.

All points in

> Quadrant I have both coordinates positive.
> Quadrant II have negative x-coordinates and positive y-coordinates.
> Quadrant III have both coordinates negative.
> Quadrant IV have positive x-coordinates and negative y-coordinates.

The points on the axes themselves are not considered in any quadrant. All the points on the x-axis have coordinates of the form $(x, 0)$; all the points on the y-axis have coordinates of the form $(0, y)$.

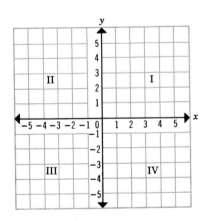

Figure 6.9

EXERCISES 6.2

In which quadrant does each of the points corresponding to the ordered pairs below belong (Exercises 1–10)?

1. $(3, 7)$
2. $(-7, 8)$
3. $(3, -8)$
4. $(1.7, -2.8)$
5. $(-\frac{1}{2}, -\frac{1}{4})$
6. $(0.0003, -0.0004)$
7. $(-17\frac{1}{4}, -17\frac{1}{2})$
8. $(-3.86, 1.92)$
9. $(4.874, -111.1)$
10. $(3.04, 1.076)$

11. What is the x-coordinate of all the points on the y-axis?
12. What is the y-coordinate of all the points on the x-axis?
13. What can you say about the signs of the coordinates of a point P, with label (x, y), if it is in:
    (a) The first quadrant?
    (b) The second quadrant?
    (c) The third quadrant?
    (d) The fourth quadrant?
14. Where is a point if its abscissa is positive?
15. Where is a point if its ordinate is positive?
16. Where is a point if its abscissa is negative?
17. Where is a point if its ordinate is negative?
18. In which quadrant is each of the following points if n is a positive integer?
    (a) $(5, n)$
    (b) $(n, 6)$
    (c) $(-n, -2)$
    (d) $(-n, 5)$
    (e) $(n^2, -n)$
    (f) $(-n^2, n)$
    (g) $(n + 2, n^2)$
    (h) $(-n, -n)$
19. Plot the points with the following labels.
    (a) $(-3, 2)$
    (b) $(0, 4)$
    (c) $(0, -6)$
    (d) $(-3, 0)$
    (e) $(-4, -4)$
    (f) $(5, 0)$
    (g) $(3, 2)$
    (h) $(5, -3)$
    (i) $(0, 0)$
20. Where are all the points with x-coordinate 4?

6.3 ◆ Open Sentences in Two Variables

Open sentences such as

$$y = x + 3$$
$$x + y = 8$$
$$2x + y = 7$$
$$x + 7 < y$$

are called **open sentences containing two variables.** Let us use the set

$$U = \{1, 2, 3, 4, 5, 6, 7, 8, 9\}$$

as the replacement set for both of the variables. We want to find all ordered pairs of the form (x, y), where x and y are elements of U, that have the property whereby replacement of x by the first component and replacement of y by the second component in the above open sentences produces a true statement. For example, $(1, 4)$ is an ordered pair with the property that, if x is replaced by 1 and y by 4 in the open sentence $y = x + 3$, a true statement is produced. An ordered pair for which this occurs is called a **solution** of the open sentence in two variables, and the set of all such ordered pairs is called the **solution set** of the sentence. Each ordered pair in the solution set is said to **satisfy** the open sentence.

To find the ordered pairs that are solutions of a given equation or inequality, we can assign values to one of the variables and then determine the paired value of the other. Thus, for the equation $y = x + 3$, if we assign any value to x from the given replacement set, we can find the paired value of y by adding 3 to the assigned value of x. For example, if we assign 2 as a value of x, we see that 5 is the paired value of y and that $(2, 5)$ is a solution of this equation.

In finding ordered pairs that satisfy a given equation or inequality, it is convenient to find an equivalent equation or inequality that directly expresses a relation giving one variable in terms of the other. For example, before assigning values to x in

$$y + 2x < 5$$

we transform this inequality into the equivalent inequality

$$y < 5 - 2x$$

In the second of these inequalities we say that the two variables are **explicitly** related—that is, we have one member of the inequality that consists of one of the variables standing alone and one member consisting of some expression involving the other variable.

If the replacement set of the variables x and y is

$$U = \{1, 2, 3, 4, 5, 6, 7, 8, 9\}$$

then the solution set of the equation

$$y = x + 3$$

is

$$\{(1, 4), (2, 5), (3, 6), (4, 7), (5, 8), (6, 9)\}$$

With the same replacement set, the solution set of the equation

$$x + y = 8$$

is
$$\{(1, 7), (2, 6), (3, 5), (4, 4), (5, 3), (6, 2), (7, 1)\}$$

Again, with the same replacement set, the solution set of

$$2x + y = 7$$

is

$$\{(1, 5), (2, 3), (3, 1)\}$$

The solution set of

$$x + 5 < y$$

is

$$\{(1, 7), (1, 8), (1, 9), (2, 8), (2, 9), (3, 9)\}$$

(using the same replacement set). The graphs of these solution sets are shown in Figure 6.10.

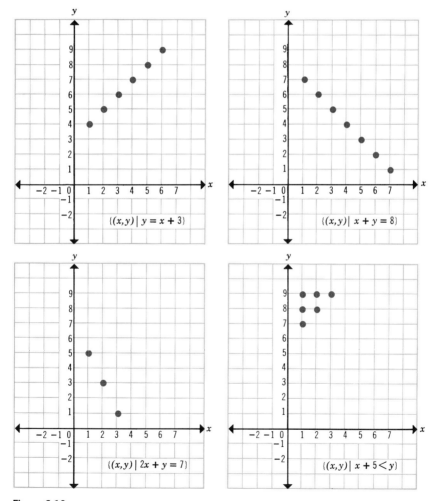

Figure 6.10

If we choose as the replacement set of the variables in the open sentences above the set of all real numbers, then the solution sets are given below with their graphs (Figure 6.11).

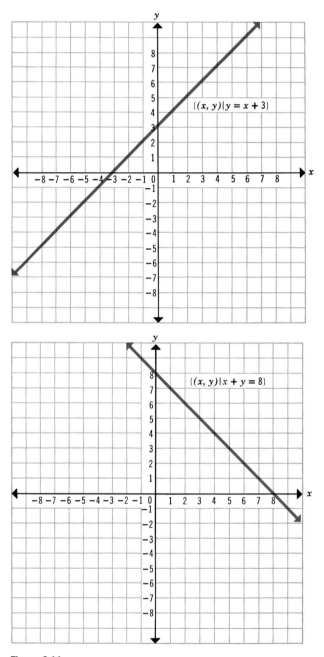

Figure 6.11

6.3 OPEN SENTENCES IN TWO VARIABLES

Notice that, on the graph of $\{(x, y) \mid x + 5 < y\}$, every point on the shaded side of the straight line belongs to the graph of the solution set but that the points on the line itself do not belong to the graph. If the line itself were included we would indicate this by darkening it.

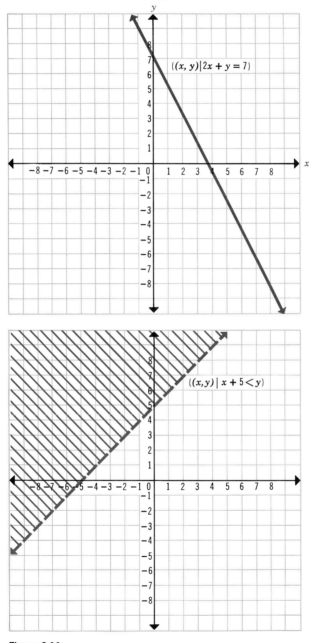

Figure 6.11

6.4 ✦ Relations and Functions

Given a set A, the set $A \times A$ is the set of all possible ordered pairs with components from A. Given the sets A and B, the set $A \times B$ is the set of all possible ordered pairs with first components from A and second components from B. In this book we shall study sets of ordered pairs of real numbers. Any set of ordered pairs is called a **relation**. The sets below are relations:

$$\{(0, 0), (0, 2), (0, 3), (0, -4)\}$$
$$\{(1, 2), (2, 3), (3, 4), (4, 5)\}$$
$$\{(1, 2), (2, 4), (3, 6), (4, 8), (5, 10), (6, 12)\}$$

The **domain** of a relation is the set of all the first components of the ordered pairs that make up that relation. The **range** of a relation is the set of all the second components of the ordered pairs that make up that relation. The domain of the first relation above is $\{0\}$, and its range is $\{0, 2, 3, -4\}$. In the second relation, the domain is $\{1, 2, 3, 4\}$ and the range is $\{2, 3, 4, 5\}$.

Relations are usually given in one of three ways. One way is simply to list the ordered pairs that belong to a relation. This is called the **roster** method. Some relations are given by the roster method in the examples below.

EXAMPLE 1

$$\{(1, 2), (2, 3), (3, 4)\}$$

EXAMPLE 2

$$\{(0, 0), (0, -1), (0, -2), (-1, -2)\}$$

Another way is to display the graph of the relation—that is, the set of those points in the Cartesian plane which have as their labels the ordered pairs of the given relation. The graphs of the above examples are shown in Figure 6.12.

A third method of designating a relation is by means of the solution set of an open sentence in two variables. For example, consider the open sentence $y = -x$. The solution set of this sentence, namely, $\{(x, y) \mid y = -x\}$, is a set of ordered pairs and, therefore, is a relation. The solution set of every open sentence in two variables is a relation.

Consider the relations given by the graphs in Figure 6.13. Notice that any vertical line intersects the graphs in Figure 6.13a and Figure 6.13c in at most one point. We see then, that no two ordered pairs of these relations have the same first components and different second components.

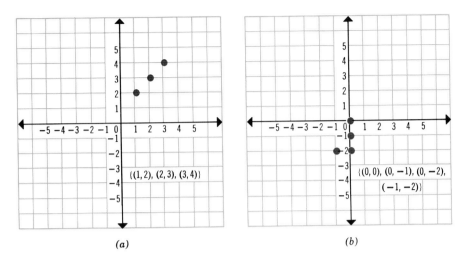

Figure 6.12 (a) Example 1. (b) Example 2.

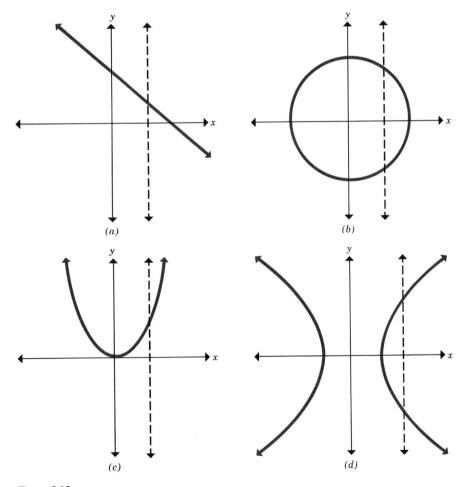

Figure 6.13

On the other hand, in Figure 6.13b and Figure 6.13d, we see that some vertical lines intersect the graph in more than one point. Some ordered pairs of these relations have the same first components and different second components.

The relations given in Figure 6.13a and Figure 6.13c are called functions. A **function** is a relation no two of whose ordered pairs have the same first component and different second components. The relations in Figure 6.13b and Figure 6.13d are not functions. Since a function is a special kind of relation, we see that every function is a relation but not every relation is a function.

It is important to determine whether a relation, regardless of how it is specified, is a function. If a relation is designated by the roster method, the decision is easy. For example, the relation

$$\{(1, 1), (1, 2), (3, 4), (5, 6)\}$$

is not a function, since two of its ordered pairs, $(1, 1)$ and $(1, 2)$, have the same first component and different second components. On the other hand, the relation

$$\{(2, 4), (3, 6), (4, 8), (6, 12)\}$$

is a function, since no two of its ordered pairs have the same first components. It is also easy to decide whether relations specified by graphs are functions. Observe, in Figure 6.14, that there is at least one vertical line that intersects the graph of the relation in more than one point; hence this relation contains two or more ordered pairs with the same first component and is not a function.

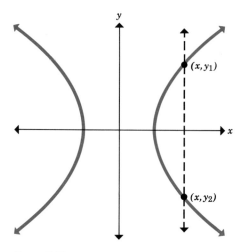

Figure 6.14

6.4 RELATIONS AND FUNCTIONS

Determining whether or not a relation is a function using a vertical line as we did above is called the **vertical line test** for a function.

Now let us consider the case where the relation is specified by an open sentence in two variables. If, for any replacement of the variable which is the first component, there are two or more replacements for the variable which is the second component, then the open sentence *does not* specify a function. For example:

1. $\{(x, y) \mid y = x - 3\}$ is a function because, for any particular replacement for x, only one value of y will satisfy $y = x - 3$.

2. $\{(x, y) \mid x^2 + y^2 = 4\}$ is *not* a function because, for some replacements for x, there are two values of y that satisfy $x^2 + y^2 = 4$. For example, $(0, 2)$ and $(0, -2)$ will satisfy this equation.

EXERCISES 6.3

Find the missing components so each ordered pair will satisfy the given equation (Exercises 1–10).

1. $y = x - 7$ (3,)
2. $y = 3x + 6$ (0,)
3. $y = 2x + 7$ (, −3)
4. $x + 3y = 15$ (, 3)
5. $2x - y = 12$ (, 0)
6. $y^2 = x + 3$ (, 4)
7. $y = x^2$ (−1,)
8. $xy = 15$ (−3,)
9. $xy + 7 = 21$ (2,)
10. $x^2 + 4y = 3$ (2,)

Graph the following relations (Exercises 11–20).

11. $\{(-6, 5), (-6, 4), (3, -2), (1, 0)\}$
12. $\{(x, y) \mid y = x - 2\}$
13. $\{(x, y) \mid y = 3x\}$
14. $\{(x, y) \mid x = 0\}$
15. $\{(x, y) \mid 2x - 3y = 6\}$
16. $\{(x, y) \mid x < y\}$
17. $\{(x, y) \mid y > 2 + x\}$
18. $\{(x, y) \mid y = x^2\}$
19. $\{(x, y) \mid x + y > 4\}$
20. $\{(x, y) \mid |x| = |y|\}$

21. If R is the set of all real numbers, the graph of $R \times R$ is the entire plane. For each graph below (*a* to *h*), tell whether it is the graph of a relation or of a function.

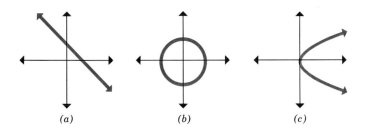

(a)   (b)   (c)

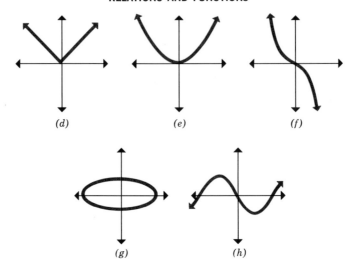

State the domain and range of each of the following relations (Exercises 22–25).

22. $\{(-3, 4), (-5, 12), (2, 9), (0, 7)\}$
23. $\{(1, 1), (2, 2), (3, 3), (4, 4)\}$
24. $\{(6, 7)\}$
25. $\{\ldots, (-2, 0), (-1, 0), (0, 0), (1, 0), \ldots\}$

Are the following sentences true or false (Exercises 26–35)?

26. $\{(-4, 1), (-3, 2), (-2, 0), (-1, 1)\}$ is a function.
27. $\{(-3, -2), (3, -1), (3, 0), (3, 2)\}$ is a function.
28. $\{(1, 0), (2, 0), (3, 0), (4, 0)\}$ is a function.
29. $\{(x, y) \mid y = x^2\}$ is a function.
30. $\{(x, y) \mid y = 2x - 4\}$ is a function.
31. $\{(x, y) \mid x^2 + y^2 = 4\}$ is a function.
32. $\{(x, y) \mid y = x^3\}$ is a function.
33. The domain of the relation $\{(0, 1), (0, 2), (0, 3), (0, 4)\}$ is $\{0\}$.
34. The range of the relation $\{(3, 1), (3, 2), (3, 3), (3, 4)\}$ is $\{3\}$.
35. The range of the relation $\{(1, 4), (2, 5), (3, 6), (5, 8)\}$ is $\{4, 5, 6, 8\}$.

6.5 ◆ Functional Notation

A function is usually named by the use of letters $f, g, \ldots$ or some other lowercase letter. For example, f may be the set of ordered pairs (x, y) for which $y = 3x$. We write this

$$f = \{(x, y) \mid y = 3x\}$$

The open sentence $y = 3x$ specifies the function. With this open sentence we can decide whether or not any given ordered pair belongs to the function. Thus $(1, 3)$ is an ordered pair of this function and $(2, 3)$ is not.

6.5 FUNCTIONAL NOTATION

Frequently, instead of writing (x, y) for one of the ordered pairs of a function, f, the symbol, $(x, f(x))$ is used. The function f will then be denoted by

$$f = \{(x, f(x)) \mid f(x) = 3x\}$$

The symbol $f(x)$, which is read "**f at x**" or "**the value of f at x**," is used to designate the second component of the ordered pair of f whose first component is x.

We call a second component of an ordered pair belonging to a function a **value** of the function. For example, for the function given above, the value of f at 1, denoted by $f(1)$, is 3; the value of f at 2, denoted by $f(2)$, is 6.

EXAMPLE. Let $g = \{(x, g(x)) \mid g(x) = 2x - 3\}$. Find $g(0)$, $g(1)$, and $g(-2)$.

SOLUTION

$$g(0) = 2(0) - 3 = -3$$
$$g(1) = 2(1) - 3 = -1$$
$$g(-2) = 2(-2) - 3 = -7$$

EXERCISES 6.4

Each of Exercises 1–5 refers to one of these five functions:

$$A = \{(-3, 4), (-2, 3), (2, 3), (3, 4)\}$$
$$B = \{(-3, -2), (-2, 0), (-1, -2), (0, 4), (1, 2), (2, 1)\}$$
$$f = \{(x, f(x)) \mid f(x) = x^2\}$$
$$g = \{(x, g(x)) \mid g(x) = 5x - 1\}$$
$$h = \{(x, h(x)) \mid h(x) = 2 - 3x\}$$

1. True or false.
   (a) $A(-3) = 4$
   (b) $f(-1) = 1$
   (c) $h(-2) = -4$
   (d) $g(0) = 1$
   (e) $B(0) = -2$
   (f) $g(1) = 4$
   (g) $f(-2) = -4$
   (h) $g(f(2)) = 19$
2. Find the value of each of the following.
   (a) $f(-3)$
   (b) $A(3)$
   (c) $B(0)$
   (d) $h(-2)$
   (e) $A(2)$
   (f) $g(-1)$
   (g) $f(-4)$
   (h) $h(0)$
   (i) $B(g(0))$
   (j) $f(g(-1))$
3. What are the domains of the functions A, B, h, and f?
4. What are the ranges of the functions f, g, h, and A?
5. Draw the graph of the function g.

6.6 • Linear Functions and Their Graphs

A function f is a **linear function** if and only if there are real numbers, a and b, with $a \neq 0$, such that

$$f = \{(x, y) \mid y = ax + b\}$$

It can be proved that the graph of a linear function is a straight line.

Since the graph of a linear function is a straight line, and a straight line is completely determined by two points, a linear function is completely determined by two of its ordered pairs. In practice, the two ordered pairs most easily found are those with first or second components zero—that is, ordered pairs of the form $(x, 0)$ and $(0, y)$.

Replacing y by 0 in the open sentence $y = ax + b$ gives

$$0 = ax + b$$

which is equivalent to

$$x = -\frac{b}{a}$$

Since $a \neq 0$, $\left(-\frac{b}{a}, 0\right)$ is an ordered pair of this function. Similarly, if x is replaced by 0 in the open sentence $y = ax + b$, we have

$$0 + b = y$$
$$y = b$$

and $(0, b)$ is another ordered pair of this function. Since the points with these labels are the points where the graph crosses the x-axis and the y-axis, they are easy to locate. The number $-\frac{b}{a}$ is called the **x-intercept** of the graph and the number b is called the **y-intercept.**

We see then that in the equation $y = ax + b$, the number b is the y-intercept of the line which is the graph of the equation.

EXAMPLE 1. Given $f = \{(x, y) \mid y = 2x + 6\}$, find the x and y-intercepts.

SOLUTION. In the equation $y = 2x + 6$, which is of the form $y = ax + b$, we see that the number $b = 6$ is the y-intercept. Since the y-intercept is 6, the point $(0, 6)$ is on the graph. Replacing y by 0 in the given equation, we have

$$2x + 6 = 0$$
$$x = -3$$

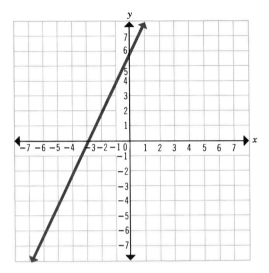

Figure 6.15

and the x-intercept is -3. Thus $(-3, 0)$ is on the graph of $y = 2x + 6$. The graph of the function

$$\{(x, y) \mid y = 2x + 6\}$$

which is the graph of the open sentence $y = 2x + 6$, is shown in Figure 6.15.

It may happen that the graph intersects the axes at the origin, so that the intercepts are not different numbers and do not represent different points. In this case it is necessary to determine another point in order to draw the graph. It is advisable in all cases to determine a third point as a check.

Given an equation that specifies a linear function, we can find the slope of the graph of the function. The **slope** of a line is the ratio of the rise to the run of any segment of the line (Figure 6.16).

If P_1 with coordinates (x_1, y_1) and P_2 with coordinates (x_2, y_2) are two points on a line, then the **slope**, which is usually denoted by m, is given by

$$m = \frac{y_2 - y_1}{x_2 - x_1} \qquad x_2 \neq x_1$$

For example, if the points $(1, 3)$ and $(-3, -1)$ are on a line, we have

$$m = \frac{y_2 - y_1}{x_2 - x_1} = \frac{(-1) - 3}{(-3) - 1} = \frac{-4}{-4} = 1 \text{ (see Figure 6.17)}$$

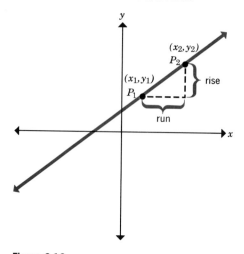

Figure 6.16

Now let us consider an equation of a line passing through a given point on the y-axis whose coordinate is $(0, b)$ and which has a slope m. Let P be any point on the line. Let the coordinates of P be (x, y) $x \neq 0$. Using the formula for the slope, we have

$$m = \frac{y - b}{x - 0}$$

from which we obtain

$$y = mx + b$$

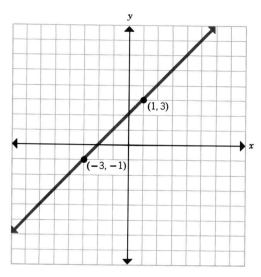

Figure 6.17

6.6 LINEAR FUNCTIONS AND THEIR GRAPHS

where the number m is the slope and the number b is the y-intercept of the line.

We see then that $y = mx + b$ is an equation whose graph is a straight line with slope m and y-intercept b.

EXAMPLE 2. Draw the graph of the function $g = \{(x, y) \mid y = -2x + 4\}$, and give the slope and the y-intercept of the graph.

SOLUTION. Since $y = -2x + 4$ is of the form $y = mx + b$ we see that the slope of the line is -2 and the y-intercept is 4. Since the y-intercept is 4, the point $(0, 4)$ is on the graph. Replacing y by 0 in $y = -2x + 4$, we see that

$$0 = -2x + 4$$
$$x = 2$$

so that $(2, 0)$ is on the graph. The graph of the function is shown in Figure 6.18.

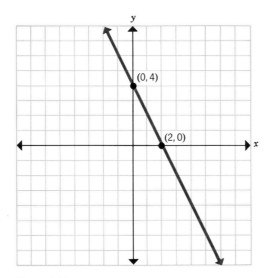

Figure 6.18

EXERCISES 6.5

1. Which of the following equations specifies a linear function?
   (a) $f(x) = 3x - 6$
   (b) $f(x) = x^2 + 2$
   (c) $f(x) = x - 5$
   (d) $f(x) = 2 - 3x^2$
   (e) $f(x) = -3 + 7x$
   (f) $f(x) = 2 - 3x$

2. Determine the x and y-intercepts of the linear functions given by the following equations.
   (a) $y = 3x + 6$
   (b) $y = \frac{1}{2}x + 4$
   (c) $y = 4 - 6x$
   (d) $x + 3y = 9$
   (*Hint.* First find an equivalent equation of the form $y = mx + b$.)
   (e) $x + y = 0$
   (f) $\frac{1}{2}y = 3x + 2$
   (g) $4x - 3y = 12$
   (h) $3 - 2y + 7x = 0$
   (i) $3y - 4x = 2x + 6y - 4$

Graph the linear functions specified by the equations below (Exercises 3–10).

3. $y = 2x + 4$
4. $y = 3x + 6$
5. $y = 2x + 1$
6. $y = -3x + 6$
7. $x + y = 7$
8. $2x - 3y = 18$
9. $4x = y - 6$
10. $x + 3y = 9$

11. We observe that in $y = mx + b$, m is the slope of the graph. Give the slope of the graph of each of the following.
    (a) $y = 3x + 6$
    (b) $y = -2x - 7$
    (c) $y = \frac{1}{2}x + 4$
    (d) $x + y = 3$
    (e) $2y - 3x = 4$
    (f) $4x + 2y = 2$
    (g) $7x - 6y = 18$
    (h) $2x + y = -7$

12. Each pair of points below determines a line. Give the slope of each line determined.
    (a) $(2, 1), (1, 3)$
    (b) $(4, -1), (3, 2)$
    (c) $(-1, -1), (4, 1)$
    (d) $(0, 3), (4, -2)$
    (e) $(0, 0), (-1, -1)$
    (f) $(7, 0), (3, 8)$
    (g) $(1, 2), (2, 2)$
    (h) $(3, -1), (-3, -2)$

6.7 ◆ Direct and Inverse Variation

The function

$$f = \{(x, y) \mid y = kx\}$$

where k (called the **constant of variation**) is a non zero constant, is an example of **direct variation**. The variable y in the equation $y = kx$ is said to **vary directly** as the variable x. Some examples of such variations are given by familiar formulas. For example, in the formula for finding the circumference of a circle

$$C = 2\pi r$$

6.7 DIRECT AND INVERSE VARIATION

the circumference, C, varies directly as the radius, r. Here the constant of variation is 2π.

Another situation in which the phrase "direct variation" is used occurs in

$$y = kx^2$$

where k is a nonzero constant. In this case we say that the variable y varies directly as the square of the variable x. For example, the formula

$$A = \pi r^2$$

asserts that the area, A, of a circle varies directly as the square of the radius, r. Here the constant of variation is π.

A second type of variation arises from the equation

$$xy = k$$

where k is a nonzero constant and $x \neq 0$. In this case we say that x and y **vary inversely**. If we write the equation $xy = k$ in a form that specifies a function, we have

$$y = \frac{k}{x}$$

and y is said to **vary inversely** as x. As an example, consider the set of all rectangles with area of 12 square units. Since the area of a rectangle is given by

$$A = LW$$

and we have the equation

$$12 = LW$$

That is, the length, L, and the width, W, vary inversely.

If we know that two variables vary directly or inversely and we have an ordered pair of values of these variables, we can find the constant of variation. For example, suppose y varies directly as x and when x is 2 then y is 6. Since y varies directly as x, we know that there is some nonzero constant, k, such that

$$y = kx$$

The fact that y is 6 when x is 2 tells us that $(2, 6)$ is a solution of $y = kx$ and hence that

$$6 = 2k$$

from which we know that k is 3. The equation specifying this direct variation then is

$$y = 3x$$

As another example, if we know that y varies inversely as x and that y is 4 when x is 3, then we have

$$(3)(4) = k$$

so that k is 12 and x and y are related by the equation $xy = 12$.

When several variables are related, we say that each of them varies **jointly** with the others. This variation may be direct, or inverse, or a combination. Thus, if y varies jointly with x and z, and directly as x and z, then there is some nonzero constant, k, such that

$$y = kxz$$

If y varies jointly with x and z, directly as x and inversely as z, we have some nonzero constant, k, such that

$$y = k\frac{x}{z},\ z \neq 0$$

Some relations are also called **proportions.** The word proportion arises from the fact that any two solutions (a, b) and (c, d) of an equation expressing a direct variation, $y = kx$, satisfy an equation of the form

$$\frac{a}{b} = \frac{c}{d}$$

(since both $b = ka$ and $d = kc$ must be true). Such an equation is called a proportion. For example, consider the situation in which the resistance, R, of a wire varies directly as the length, L, and inversely as the square of the diameter, d. This can be represented by the equation

$$R = k\frac{L}{d^2}$$

For a given set of values of these variables: L_1, R_1, and d_1, we have

$$R_1 = k\frac{L_1}{d_1^2} \quad \text{or} \quad k = \frac{R_1 d_1^2}{L_1}$$

For a second set of values of these variables: R_2, L_2, and d_2, we have

$$R_2 = k\frac{L_2}{d_2^2} \quad \text{or} \quad k = \frac{R_2 d_2^2}{L_2}$$

6.7 DIRECT AND INVERSE VARIATION

Hence
$$\frac{R_1 d_1^2}{L_1} = \frac{R_2 d_2^2}{L_2}$$
and the value of any variable can be determined if the values of the other variables are known.

EXERCISES 6.6

Write an equation expressing the relationship between the variables, using k as the constant of variation (Exercises 1–15).

1. The distance, d, traveled by a car moving at a constant rate varies directly as the time, t.
2. The elongation, E, of a spring balance varies directly as the applied weight, W.
3. The amount, A, of money earned varies directly as the number, h, of hours worked.
4. The weight, W, of a model dam varies directly as the cube of its height, H.
5. The kinetic energy, K, of a moving body varies directly as the square of its velocity, V.
6. If the volume of a gas remains constant, the pressure, P, varies directly as the absolute temperature, T.
7. If the temperature of a gas remains constant, the pressure, P, varies inversely as the volume, V.
8. The force, F, with which the earth attracts an object above the earth's surface varies inversely as the square of the distance, D, from the center of the earth.
9. The perimeter, P, of a square varies directly as the length of a side, S.
10. The distance, d, that a falling object will drop varies directly as the square of the time, t.
11. The intensity, B, of light varies inversely as the square of the distance, d, from the source.
12. The force, F, of the wind blowing directly upon a vertical plane varies jointly as the area, A, of the surface and the square of the velocity, v.
13. The horsepower, H, that a rotating shaft can safely transmit varies jointly as the cube of its diameter, d, and the number, N, of revolutions it makes per minute.
14. The strength, S, of a horizontal beam of rectangular cross section and of length, L, when supported at both ends varies jointly as the breadth, b, and the square of the depth, d, and inversely as the length, L.
15. The force, F, of attraction between two spheres varies directly as the product of their masses, m_1 and m_2, and inversely as the square of the distance, d, between their centers.

16. If y varies directly as x, and y is 15 when x is 20, what is the value of y when x is 9?
17. If y varies directly as x^2 and y is 72 when x is 3, what is the value of y when x is 7?
18. If y varies directly as x^3 and y is 8 when x is 2, what is y when x is 6?
19. If y varies jointly as x and z, and y is 5 when x is 6 and z is 7, find y when x is 3 and z is 14.
20. If y varies inversely as x and y is 6 when x is 4, find y when x is 12.
21. A pressure, P, of 1830 pounds per square foot is exerted by 2 cubic feet of air in a cylinder fitted with a piston. If the piston is pushed out until the pressure becomes 1220 pounds per square foot, what will be the volume of air if there is no change in temperature? (See Exercise 7.)
22. The areas of two similar figures are directly proportional to the squares of the corresponding sides. If two similar triangles have corresponding sides of 5 feet and 10 feet, and if the area of the first triangle is 125 square feet, what is the area of the second triangle?
23. The distance, d, in miles that a person can see to the horizon from a point h feet above the surface of the earth varies as the square root of the height, h. If for a height of 144 feet the horizon is 18 miles distant, how far is the horizon from a point that is 900 feet high?
24. The electrical resistant, R, of a wire of uniform cross section varies directly as its length, L, and inversely as its cross-sectional area, A. A solid cube of copper 1 inch on a side has a resistant of $6.85 \ (10)^{-7}$ ohm at $20°C$. What is the resistance of a copper wire 100 feet long and 0.04 inches in diameter at the same temperature?

CHAPTER 6 REVIEW

1. If $A = \{-2, -3, -4\}$, list the elements of $A \times A$.
2. $B = \{-2, 0, 2\}$. Graph $B \times B$.
3. In which quadrant is each of the following points (n is a positive integer)?
   (a) $(n, 2)$ (d) $(n^2, -n)$
   (b) $(3, -n)$ (e) $(-n^2, n^2)$
   (c) $(-n, n)$ (f) $(n, -n - 2)$
4. Graph the relation $\{(x, y) \mid y + x = 3\}$.
5. State the domain and range of the relation
$$\{(-2, 3), (-3, 6), (-4, 9), (4, 7)\}$$
6. If $f = \{(x, f(x)) \mid f(x) = x^2\}$, find:
   (a) $f(0)$ (c) $f(-1)$
   (b) $f(4)$ (d) $f(-4)$

CHAPTER 6 REVIEW

Given $f(x) = x + 2$ and $g(x) = x^3 - 3$, find the following (Exercises 7–12).

7. $f(-2)$
8. $g(-2)$
9. $\dfrac{f(2)}{g(3)}$
10. $[f(2)]^2$
11. $f(g(3))$
12. $g(x + h) - g(x)$

13. Determine the x and y-intercepts of the graph of the linear function given by $y = 3x - 9$.
14. Graph the linear function specified by $y = -3x + 9$.
15. What is the slope of the line which is the graph of $y = \frac{3}{2}x - 7$?
16. Give the slope of the line determined by the two points whose coordinates are $(-3, -4)$ and $(2, -5)$.
17. The tension, T, on a spring varies directly as the distance, s, that it is stretched. Write an equation expressing the relationship between the variables, using k as the constant of variation.
18. The force of attraction, F, between two bodies varies directly as the product of their masses, m and M, and inversely as the square of the distance, d, between them. Write an equation expressing this relation, using k as the constant of variation.
19. Graph the function specified by the equation $x + 3y = 6$. What are the x- and y-intercepts? What is the slope?
20. Graph the relation

$$\{(x, y) \mid y < 2x - 3\}$$

CHAPTER 7

Complex Numbers

7.1 • Extension of the Real Number System

There are many equations involving polynomials over the real numbers that do not have solutions in the set of real numbers. For instance, we are unable to find real numbers that are solutions of the equation $x^2 = -1$, since the laws governing the real number system require that the square of a real number must be non negative. In this chapter we shall develop a new number system, called the system of **complex numbers,** which will enable us to solve such equations.

A **complex number** is defined as an ordered pair of real numbers. Some examples of complex numbers are

$$(3, 2), (0, -1), (\sqrt{2}, -\sqrt{2}), (1, -\sqrt{3})$$

We shall designate the set of all complex numbers by the letter C. For a complex number of the form (a, b)—where a and b are real numbers—the real number a is called the **real component,** or **real part** of (a, b), and the real number b is called the **imaginary component,** or **imaginary part** of (a, b). The words "real" and "imaginary" are used for historical reasons and not in their usual English context.

Two complex numbers $z_1 = (a, b)$ and $z_2 = (c, d)$ are said to be **equal** if and only if $a = c$ and $b = d$. Two complex numbers are equal then if and only if their real components are equal and their imaginary components are equal. Thus

$(\sqrt{4}, \sqrt{9}) = (2, 3)$ since $\sqrt{4} = 2$ and $\sqrt{9} = 3$
$(3 + 4, -1) = (7, -1)$ since $3 + 4 = 7$ and $-1 = -1$

We now define the operations of addition, subtraction, multiplication, and division for the complex numbers, using the operations of the real num-

7.2 ♦ Addition of Complex Numbers

We define the addition of two complex numbers,

$$z_1 = (a, b) \quad \text{and} \quad z_2 = (c, d)$$

by

$$z_1 + z_2 = (a, b) + (c, d) = (a + c, b + d)$$

Note that we used the addition of real numbers to define the addition of complex numbers, since the real components, a and c, and the imaginary components, b and d, are real numbers.

EXAMPLES

$$(3, 4) + (-3, 2) = (3 + [-3], 4 + 2) = (0, 6)$$
$$(8, -4) + (-3, -2) = (8 + [-3], -4 + [-2]) = (5, -6)$$
$$(\sqrt{2}, \tfrac{1}{2}) + (\sqrt{3}, -\tfrac{1}{4}) = (\sqrt{2} + \sqrt{3}, \tfrac{1}{2} + [-\tfrac{1}{4}]) = (\sqrt{2} + \sqrt{3}, \tfrac{1}{4})$$

The set of complex numbers is closed under this operation of addition, since if $z_1 = (a, b)$ and $z_2 = (c, d)$, then

$$z_1 + z_2 = (a, b) + (c, d)$$
$$= (a + c, b + d) \quad \text{definition of addition of complex numbers}$$

But a, b, c, and d are real numbers, hence $a + c$ and $b + d$ are real numbers by the closure property of addition of real numbers. Since $a + c$ and $b + d$ are real numbers, $(a + c, b + d)$ is a complex number by definition of a complex number.

Addition of complex numbers is a commutative operation. Let $z_1 = (a, b)$ and $z_2 = (c, d)$. Then

$$z_1 + z_2 = (a, b) + (c, d)$$
$$= (a + c, b + d) \quad \text{definition of addition of complex numbers}$$
$$= (c + a, d + b) \quad \text{commutative property of addition of real numbers}$$
$$= (c, d) + (a, b) \quad \text{definition of addition of complex numbers}$$
$$= z_2 + z_1$$

We now show that addition of complex numbers is an associative operation. Let $z_1 = (a, b)$, $z_2 = (c, d)$, and $z_3 = (e, f)$. Then

$$(z_1 + z_2) + z_3 = [(a, b) + (c, d)] + (e, f)$$
$$= (a + c, b + d) + (e, f) \quad \text{definition of addition of complex numbers}$$
$$= ([a + c] + e, [b + d] + f) \quad \text{definition of addition of complex numbers}$$
$$= (a + [c + e], b + [d + f]) \quad \text{associative property of addition of real numbers}$$
$$= (a, b) + (c + e, d + f) \quad \text{definition of addition of complex numbers}$$
$$= (a, b) + [(c, d) + (e, f)] \quad \text{definition of addition of complex numbers}$$
$$= z_1 + (z_2 + z_3)$$

The complex number $(0, 0)$ is the **additive identity** of the set of complex numbers, since
$$(a, b) + (0, 0) = (a + 0, b + 0) = (a, b)$$

Every complex number, (a, b), has as its **additive inverse** the complex number $(-a, -b)$, since
$$(a, b) + (-a, -b) = (a + [-a], b + [-b]) = (0, 0)$$

We will denote the additive inverse of the complex number $z = (a, b)$, by
$$-z = (-a, -b) = -(a, b)$$

7.3 ◆ Subtraction of Complex Numbers

In the set of real numbers, we defined the difference, $a - b$, as that real number c such that $b + c = a$. Similarly, in the set of complex numbers, if $z_1 = (a, b)$ and $z_2 = (c, d)$, we define the difference, $z_1 - z_2$, as that complex number z_3 such that $z_2 + z_3 = z_1$. We shall now show that
$$z_1 - z_2 = z_1 + (-z_2)$$

By the definition of subtraction for complex numbers, we know that if $z_1 - z_2 = z_3 = (x, y)$, then
$$z_2 + z_3 = z_1$$
$$(c, d) + (x, y) = (a, b)$$
$$(c + x, d + y) = (a, b)$$

From the definition of equality for complex numbers, we have
$$c + x = a \quad \text{and} \quad d + y = b$$
$$x = a - c \quad \text{and} \quad y = b - d$$

7.3 SUBTRACTION OF COMPLEX NUMBERS

Thus

$$
\begin{aligned}
z_3 = z_1 - z_2 &= (a - c, b - d) \\
&= (a + [-c], b + [-d]) \quad \text{definition of subtraction of real numbers} \\
&= (a, b) + (-c, -d) \quad \text{definition of addition for complex numbers} \\
&= (a, b) + [-(c, d)] \quad \text{definition of additive inverse for complex numbers} \\
&= z_1 + (-z_2)
\end{aligned}
$$

EXAMPLES

$$
\begin{aligned}
(3, 4) - (2, 3) &= (3, 4) + [-(2, 3)] \\
&= (3, 4) + (-2, -3) \\
&= (3 + [-2], 4 + [-3]) \\
&= (1, 1)
\end{aligned}
$$

$$
\begin{aligned}
(6, -4) - (-3, -4) &= (6, -4) + [-(-3, -4)] \\
&= (6, -4) + (3, 4) \\
&= (9, 0)
\end{aligned}
$$

$$
\begin{aligned}
(\sqrt{2}, 4) - (\sqrt{3}, -2) &= (\sqrt{2}, 4) + [-(\sqrt{3}, -2)] \\
&= (\sqrt{2}, 4) + (-\sqrt{3}, 2) \\
&= (\sqrt{2} - \sqrt{3}, 6)
\end{aligned}
$$

EXERCISES 7.1

Determine a and b for each of the following according to the definition of equality for complex numbers (Exercises 1–10).

1. $(2, a) = (b, -3)$
2. $(-a, 6) = (5, b)$
3. $(2a, 5) = \left(6, -\dfrac{b}{2}\right)$
4. $(2 + a, 2 - b) = (4, -3)$
5. $(7 + a, 6) = (-8, 3 + b)$
6. $(a + b, a - b) = (8, 2)$
7. $(2a - b, 3a + b) = (7, 18)$
8. $(2a + b, 2a - b) = (3, 1)$
9. $(3ab, -4) = (9, 2a)$
10. $(2a + 3b, -2b) = (4, 4)$

Compute (Exercises 11–30).

11. $(8, -2) + (3, -4)$
12. $(3, 4) + (-2, -5)$
13. $(\tfrac{1}{2}, -\tfrac{1}{4}) + (\tfrac{1}{4}, -\tfrac{1}{2})$
14. $(-5, -3) + (2, 7)$
15. $(\sqrt{2}, 4) + (2\sqrt{2}, -7)$
16. $(7, 6) + (-4, -3)$
17. $(1, 0) + (0, 1)$
18. $(3, 0) + (4, 0)$
19. $(0, 5) + (6, 0)$
20. $(-2, 3) + (-7, -8)$

21. $(3, 4) - (2, 3)$
22. $(7, -3) - (2, 3)$
23. $(-2, -7) - (-5, -4)$
24. $(-3, -5) - (8, -2)$
25. $(-2, 0) - (6, 0)$
26. $(0, -2) - (0, 6)$
27. $(0, 0) - (3, 4)$
28. $(3, -3) - (4, -5)$
29. $(0, 0) - (6, -4)$
30. $(3, 7) - (8, -9)$

For each complex number below, name its additive inverse (Exercises 31–40).

31. $(2, 5)$
32. $(-3, 8)$
33. $(8, -4)$
34. $(\sqrt{5}, 2)$
35. $(0, 0)$
36. $(6, 4)$
37. $(x + y, x - y)$
38. $(\frac{1}{2}, \frac{3}{4})$
39. $(\sqrt{3}, -7)$
40. $\left(\dfrac{a}{7}, \dfrac{7-a}{8}\right)$

7.4 ◆ Multiplication of Complex Numbers

If $z_1 = (a, b)$ and $z_2 = (c, d)$ are two complex numbers, then we define their product $z_1 z_2$ by

$$z_1 z_2 = (a, b)(c, d) = (ac - bd, ad + bc)$$

EXAMPLES

$$(2, 3)(4, 1) = ([2][4] - [3][1], [2][1] + [3][4])$$
$$= (5, 14)$$
$$(-1, 3)(-2, -3) = ([-1][-2] - [3][-3], [-1][-3] + [3][-2])$$
$$= (11, -3)$$
$$(2, -3)(-2, -5) = ([2][-2] - [-3][-5], [2][-5] + [-3][-2])$$
$$= (-19, -4)$$

The set of complex numbers is closed under this operation of multiplication because the set of real numbers is closed under the operations of addition, subtraction, and multiplication.

The set of complex numbers satisfies the commutative law for multiplication, since if $z_1 = (a, b)$ and $z_2 = (c, d)$, then

$z_1 z_2 = (a, b)(c, d)$
$ = (ac - bd, ad + bc)$ definition of multiplication for complex numbers
$ = (ca - db, da + cb)$ commutative property of multiplication for real numbers
$ = (ca - db, cb + da)$ commutative property of addition for real numbers

7.4 MULTIPLICATION OF COMPLEX NUMBERS

$\qquad = (c, d)(a, b) \qquad$ definition of multiplication for complex numbers

$\qquad = z_2 z_1$

In a similar manner it can be shown that the multiplication of complex numbers is associative.

The complex number $(1, 0)$ is the **multiplicative identity** of the set of complex numbers, since for any complex number (a, b),

$(a, b)(1, 0) = ([a][1] - [b][0], [a][0] + [b][1]) \qquad$ definition of multiplication of complex numbers

$\qquad\qquad\quad = ([a][1], [b][1]) \qquad$ multiplication property of 0 in the real number system

$\qquad\qquad\quad = (a, b) \qquad$ multiplicative identity axiom of the real number system

We can also show that, in the complex number system, multiplication and addition satisfy a distributive law. If $z_1 = (a, b)$, $z_2 = (c, d)$, and $z_3 = (e, f)$, then

$z_1(z_2 + z_3) = (a, b)[(c, d) + (e, f)]$

$\qquad\qquad = (a, b)(c + e, d + f) \qquad$ definition of addition of complex numbers

$\qquad\qquad = (a[c + e] - b[d + f], a[d + f] + b[c + e])$
$\qquad\qquad\qquad\qquad\qquad\qquad$ definition of multiplication of complex numbers

$\qquad\qquad = (ac + ae - bd - bf, ad + af + bc + be)$
$\qquad\qquad\qquad\qquad\qquad\qquad$ distributive property of real numbers

$\qquad\qquad = (ac - bd + ae - bf, ad + bc + af + be)$
$\qquad\qquad\qquad\qquad\qquad\qquad$ commutative and associative properties of addition of real numbers

$\qquad\qquad = (ac - bd, ad + bc) + (ae - bf, af + be)$
$\qquad\qquad\qquad\qquad\qquad\qquad$ definition of addition of complex numbers

$\qquad\qquad = (a, b)(c, d) + (a, b)(e, f) \qquad$ definition of multiplication of complex numbers

$\qquad\qquad = z_1 z_2 + z_1 z_3$

144 COMPLEX NUMBERS

EXERCISES 7.2

Find the products.

1. $(2, 3)(1, 0)$
2. $(2, 3)(0, 1)$
3. $(2, 3)(2, -3)$
4. $(2, 3)(3, 2)$
5. $(7, -2)(1, 0)$
6. $(7, -2)(0, 1)$
7. $(7, -2)(-2, 7)$
8. $(7, -2)(-2, -7)$
9. $(-3, 4)(1, 0)$
10. $(-3, 4)(0, 1)$
11. $(-3, 4)(-3, -4)$
12. $(-3, 4)(4, -3)$
13. $(-2, 8)(-3, 7)$
14. $(2, -8)(-3, -7)$
15. $(-2, -8)(-3, 7)$
16. $(2, -8)(3, 7)$
17. $(0, 7)(3, 0)$
18. $(0, 2)(0, -4)$
19. $(4, 0)(-6, 0)$
20. $(-7, 8)(4, -9)$

7.5 ◆ Division of Complex Numbers

In the real number system we defined the quotient $\dfrac{a}{b}$ as that real number, c, such that $a = bc$. We then saw that

$$\frac{a}{b} = a\left(\frac{1}{b}\right)$$

where $\dfrac{1}{b}$ is the multiplicative inverse of b and $b \neq 0$.

We shall define division of complex numbers in the same way. That is, if z_1 and z_2 are complex numbers and $z_2 \neq (0, 0)$, then

$$\frac{z_1}{z_2} = z_1 \cdot \frac{1}{z_2}$$

Since we defined division of complex numbers in terms of the multiplicative inverse, we must derive the multiplicative inverse of the complex number z.

If $z = (a, b) \neq (0, 0)$, $\dfrac{1}{z} = (p, q)$, and if z and $\dfrac{1}{z}$ are multiplicative inverses,

$$z\left(\frac{1}{z}\right) = (1, 0)$$

$$(a, b)(p, q) = (1, 0)$$

$$(ap - bq, aq + bp) = (1, 0)$$

Then

$$ap - bq = 1 \quad \text{and} \quad aq + bp = 0$$

7.5 DIVISION OF COMPLEX NUMBERS

Since $z = (a, b) \neq (0, 0)$, we have to have either $a \neq 0$ or $b \neq 0$. In the first case, if $a \neq 0$ the equation, $ap - bq = 1$, says

$$p = \frac{bq + 1}{a}$$

Substituting this value of p into $aq + bp = 0$, we have

$$aq + b\left(\frac{bq + 1}{a}\right) = 0$$

$$a^2 q + b^2 q + b = 0$$

$$(a^2 + b^2)q = -b$$

$$q = \frac{-b}{a^2 + b^2}$$

From this we find that

$$p = \frac{1}{a}\left[b\left(\frac{-b}{a^2 + b^2}\right) + 1\right]$$

$$= \frac{1}{a} \cdot \frac{-b^2 + a^2 + b^2}{a^2 + b^2}$$

$$= \frac{1}{a} \cdot \frac{a^2}{a^2 + b^2}$$

$$= \frac{a}{a^2 + b^2}$$

On the other hand, if $b \neq 0$, a similar process again produces

$$\frac{1}{z} = \frac{1}{(a, b)} = \left(\frac{a}{a^2 + b^2}, \frac{-b}{a^2 + b^2}\right)$$

We can check that the complex number given above is indeed the multiplicative inverse of z by multiplying:

$$(a, b)\left[\frac{1}{(a, b)}\right] = (a, b)\left(\frac{a}{a^2 + b^2}, \frac{-b}{a^2 + b^2}\right)$$

$$= \left(a \cdot \frac{a}{a^2 + b^2} - b \cdot \frac{-b}{a^2 + b^2}, a \cdot \frac{-b}{a^2 + b^2} + b \cdot \frac{a}{a^2 + b^2}\right)$$

$$= \left(\frac{a^2 + b^2}{a^2 + b^2}, \frac{-ab + ab}{a^2 + b^2}\right)$$

$$= (1, 0)$$

We can now define division of one complex number by another. If z_1 and z_2 are two complex numbers, and $z_2 \neq (0, 0)$, then

$$\frac{z_1}{z_2} = z_1\left(\frac{1}{z_2}\right)$$

EXAMPLES

$$\frac{(3, 4)}{(8, 6)} = (3, 4)\frac{1}{(8, 6)}$$

$$= (3, 4)\left(\frac{8}{8^2 + 6^2}, \frac{-6}{8^2 + 6^2}\right)$$

$$= (3, 4)\left(\frac{8}{100}, -\frac{6}{100}\right)$$

$$= \left(3 \cdot \frac{8}{100} - 4\left[-\frac{6}{100}\right], 3\left[-\frac{6}{100}\right] + 4 \cdot \frac{8}{100}\right)$$

$$= \left(\frac{24 + 24}{100}, \frac{-18 + 32}{100}\right)$$

$$= \left(\frac{48}{100}, \frac{14}{100}\right)$$

$$= \left(\frac{12}{25}, \frac{7}{50}\right)$$

$$\frac{(4, 3)}{(2, 5)} = (4, 3)\frac{1}{(2, 5)}$$

$$= (4, 3)\left(\frac{2}{2^2 + 5^2}, \frac{-5}{2^2 + 5^2}\right)$$

$$= (4, 3)\left(\frac{2}{29}, \frac{-5}{29}\right)$$

$$= \left(\frac{8 + 15}{29}, \frac{-20 + 6}{29}\right)$$

$$= \left(\frac{23}{29}, \frac{-14}{29}\right)$$

EXERCISES 7.3

For each complex number, determine its multiplicative inverse (Exercises 1–10).

1. $(1, 2)$
2. $(-1, 3)$
3. $(3, -2)$
4. $(-2, -3)$
5. $(5, 0)$
6. $(0, 1)$
7. $(3, 0)$
8. $(0, -4)$
9. $(6, 4)$
10. $(-4, -9)$

Find the products (Exercises 11–20).

11. $(3, 2)(4, 3)$
12. $(7, 8)(-1, 2)$
13. $(0, 3)(3, -4)$
14. $(0, 7)(0, -2)$
15. $(4, -2)(3, -5)$
16. $(7, 2)(-4, -3)$

17. $(-2, -3)(-6, 0)$ 19. $(0, 0)(4, -3)$
18. $(1, 6)(3, 6)$ 20. $(-3, -5)(0, 0)$

Find the quotients (Exercises 21–30).

21. $\dfrac{(4, 3)}{(1, 2)}$ 26. $\dfrac{(-2, -1)}{(-4, -9)}$

22. $\dfrac{(3, 2)}{(-1, 3)}$ 27. $\dfrac{(0, 6)}{(1, 0)}$

23. $\dfrac{(2, 0)}{(-2, -3)}$ 28. $\dfrac{(10, 2)}{(-3, -2)}$

24. $\dfrac{(7, 0)}{(5, 0)}$ 29. $\dfrac{(6, 0)}{(1, -2)}$

25. $\dfrac{(4, -3)}{(0, -4)}$ 30. $\dfrac{(10, 2)}{(2, 3)}$

7.6 • Standard Form of a Complex Number

For any complex number, (a, b), we can write

$$(a, b) = (a, 0) + (0, b)$$

because of the definition of addition for complex numbers and the addition property of 0 in the real number system. Using the definition of multiplication for complex numbers, we see that

$$(a, 0) = (a, 0)(1, 0) \quad \text{and} \quad (0, b) = (b, 0)(0, 1)$$

Thus we have, for any complex number (a, b),

$$(a, b) = (a, 0)(1, 0) + (b, 0)(0, 1)$$

This means that every complex number can be expressed as the sum of products of special types of complex numbers. We have the multiplicative identity, $(1, 0)$, and the special complex number, $(0, 1)$, as factors of these products. The other factors are always complex numbers of the type $(x, 0)$, that is, complex numbers that have imaginary component 0.

When complex numbers of the form $(x, 0)$ are added, subtracted, multiplied, and divided, the following results are found:

$$(a, 0) + (b, 0) = (a + b, 0)$$
$$(a, 0) - (b, 0) = (a - b, 0)$$
$$(a, 0)(b, 0) = (ab, 0)$$
$$\dfrac{(a, 0)}{(b, 0)} = \left(\dfrac{a}{b}, 0\right) \quad \text{(where } b \neq 0\text{)}$$

As we observe the results above, a noteworthy conclusion can be drawn: *if complex numbers of the form $(a, 0)$ are subjected to the operations of addition, subtraction, multiplication, and division, the results will again be complex numbers of the same form.* This means that complex numbers of the form $(a, 0)$ behave exactly like real numbers. For this reason, complex numbers of the form $(a, 0)$ are called **purely real** complex numbers. We shall write a complex number of the form $(a, 0)$ simply as a. We see that the symbol "a" has two meanings: one as a symbol for a real number and another as a symbol for the complex number $(a, 0)$. As long as we have an expression involving only addition, subtraction, multiplication, and division, the expression will be meaningful whether we interpret the symbols in one way or another. For example, in

$$a^2 - b^2 = (a + b)(a - b)$$

the symbols a, b, $a + b$, and $a - b$ may be interpreted as symbols for real numbers or as symbols for the complex numbers $(a, 0)$, $(b, 0)$, $(a + b, 0)$, and $(a - b, 0)$. We may now represent every complex number (a, b) in the following **standard form:**

$$(a, b) = a + bi$$

where i stands for the complex number $(0, 1)$ and a and b are the complex numbers $(a, 0)$ and $(b, 0)$.

By the definition of multiplication, we have

$$i^2 = (0, 1)^2 = (0, 1)(0, 1) = (-1, 0)$$

or, with the notation adopted,

$$i^2 = -1$$

Again,

$$\begin{aligned} i^3 = i^2 \cdot i &= (0, 1)^2(0, 1) \\ &= (-1, 0)(0, 1) \\ &= (0, -1) \\ &= -(0, 1) \\ &= -i \end{aligned}$$

Observing that the fundamental laws of operation which applied to the real numbers remain true for the set of complex numbers, we conclude that, in performing fundamental operations on complex numbers represented in standard form, we can treat them as binomials, taking care always to re-

7.7 THE CONJUGATE OF A COMPLEX NUMBER

place i^2 when it appears, by -1. A few examples will show the advantages of operating with complex numbers written in standard form.

EXAMPLES

$$(1 + i)^2 = 1 + 2i + i^2 = 1 + 2i + (-1) = 2i$$
$$(3 + 2i)(3 - 2i) = 9 - 4i^2 = 9 - 4(-1) = 13$$
$$(2 - 3i)(4 + i) = 8 - 12i + 2i - 3i^2$$
$$= 8 - 10i - 3(-1) = 11 - 10i$$

7.7 ◆ The Conjugate of a Complex Number

When a complex number, $a + bi$, is written in standard form, a is called the real part and b is called the imaginary part. Two complex numbers with the same real part and opposite imaginary parts are called **conjugate**. Thus $3 + 4i$ and $3 - 4i$ are conjugate. Every complex number with zero imaginary part is its own conjugate.

The product of a complex number and its conjugate is purely real. We can prove this by taking the real numbers a and b as the real and imaginary parts of the complex number $a + bi$. The conjugate of $a + bi$ is $a - bi$, and we have

$$(a + bi)(a - bi) = a^2 - (bi)^2$$
$$= a^2 - (b^2)(i^2)$$
$$= a^2 - (b^2)(-1)$$
$$= a^2 + b^2$$

This fact is helpful in simplifying the division of complex numbers. If c and d are not both 0, we have

$$\frac{a + bi}{c + di} = \frac{(a + bi)(c - di)}{(c + di)(c - di)} = \frac{(ac + bd) + (bc - ad)i}{c^2 + d^2}$$

EXAMPLES

$$\frac{3 + 4i}{2 + i} = \frac{(3 + 4i)(2 - i)}{(2 + i)(2 - i)} = \frac{6 - 3i + 8i - 4i^2}{2^2 - i^2}$$
$$= \frac{6 + 5i - 4(-1)}{4 - (-1)}$$
$$= \frac{10 + 5i}{5}$$
$$= 2 + i$$

150   COMPLEX NUMBERS

$$\frac{2+3i}{1+2i} = \frac{(2+3i)(1-2i)}{(1+2i)(1-2i)}$$

$$= \frac{2-4i+3i-6i^2}{1^2-4i^2}$$

$$= \frac{2-i-6(-1)}{1-4(-1)}$$

$$= \frac{8-i}{5}$$

$$= \frac{8}{5} - \frac{1}{5}i$$

EXERCISES 7.4

1. Write the following complex numbers in standard form.
   (a) $(6, 3)$ (d) $(-3, -7)$ (g) $(\sqrt{2}, 2)$
   (b) $(0, 7)$ (e) $(0, -\sqrt{2})$ (h) $(\frac{1}{2}, -\frac{1}{4})$
   (c) $(4, 0)$ (f) $(-\sqrt{3}, 0)$ (i) $(3\sqrt{2}, -7)$
2. Determine x and y for each of the following to make true statements.
   (a) $3 + yi = x + 7i$ (e) $x + yi = -7i$
   (b) $-x + 6i = 4 - yi$ (f) $x + yi = 0$
   (c) $x + yi = -5 - 5i$ (g) $x + 7i = 3 - yi$
   (d) $x + yi = 6$ (h) $x + yi = 2 + 3i$
3. Give the conjugate of each of the following.
   (a) $3 + 2i$ (e) $\frac{1}{2} - \frac{1}{4}i$
   (b) $2 - 4i$ (f) $3\sqrt{2} + \sqrt{6}i$
   (c) $1 + i$ (g) $5 - 7i$
   (d) $\sqrt{2} - 2i$ (h) $6 - 2\sqrt{2}i$

Simplify (Exercises 4–29).

4. $(6 + 3i) + (4 - 2i)$ 15. $(2 + 7i) - (3 + 4i)$
5. $(7 - 3i) + (2 + 6i)$ 16. $(-1 + i) + (2 - i)$
6. $(8 + i) + (8 - i)$ 17. $(3 + i)(2 + 2i)$
7. $(2 + 3i) + (4 + 5i)$ 18. $(3 + 2i)(3 - 2i)$
8. $(-6 + 7i) + (2 - 3i)$ 19. $(5 - 3i)(5 - 3i)$
9. $(-6 + 3i) + (2 + 4i)$ 20. $(7 + 2i)(3 - 4i)$
10. $(7 + 2i) + (3 + 4i)$ 21. $(\sqrt{3} + 2i)(\sqrt{3} - 2i)$
11. $(-5 - 3i) - (-2 + 6i)$ 22. $(5 + 6i)(8 + 3i)$
12. $-(7 + 3i)$ 23. $(\sqrt{2} + 6i)(\sqrt{2} - 4i)$
13. $-(-3 - 4i)$ 24. $(-3 - \sqrt{3}i)(2 - 4\sqrt{3}i)$
14. $-(-2 + i)$ 25. $(2 + i)^2$

7.8 SQUARE ROOTS WHICH ARE COMPLEX NUMBERS

26. $(3 + 2i)^2$
27. $(\sqrt{2} + 3i)^2$
28. $(\sqrt{2} + \sqrt{3}i)^2$
29. $(6 + 8i)^2$

Simplify (Exercises 30–40).

30. $\dfrac{2 + 3i}{3 + i}$
31. $\dfrac{7 + 2i}{1 - i}$
32. $\dfrac{1 + 3i}{1 - 3i}$
33. $\dfrac{6}{4 + 3i}$
34. $\dfrac{5}{2 + 3i}$
35. $\dfrac{2 + 3i}{1 - i}$
36. $\dfrac{3 + i}{1 + 2i}$
37. $\dfrac{4 - 5i}{4 + 5i}$
38. $\dfrac{2 + \sqrt{3}i}{2 - \sqrt{3}i}$
39. $\dfrac{3 + 2i}{3 - 2i}$
40. $\dfrac{2 + 5i}{3 - 4i}$

41. Compute the following.
    (a) i^2
    (b) i^3 (Hint. $i^3 = i^2 \cdot i$.)
    (c) i^4
    (d) i^8
    (e) i^{10}
    (f) i^{16}
    (g) i^{4n} (n a positive integer)
    (h) i^{4n+1}
    (i) i^{4n+2}
    (j) i^{4n+3}

7.8 ◆ Square Roots which are Complex Numbers

Consider the following squares:

(a) $(3i)^2 = 9i^2 = -9$
(b) $(-3i)^2 = 9i^2 = -9$
(c) $(\sqrt{2}i)(\sqrt{2}i) = 2i^2 = -2$
(d) $(-\sqrt{2}i)(-\sqrt{2}i) = 2i^2 = -2$

From (a) and (b) we see that $3i$ and $-3i$ are both square roots of -9. From (c) and (d) we see that $\sqrt{2}i$ and $-\sqrt{2}i$ are both square roots of -2. Thus, if a is a negative real number, $\sqrt{-a}i$ and $-\sqrt{-a}i$ are both square roots of the complex number a. Notice that since $a < 0$ we have $-a > 0$, and $\sqrt{-a}$ is a real number. For complex numbers we do not use the concept of a principal square root, but we shall agree that for $x > 0$, when we write $\sqrt{-x}$ we mean $\sqrt{x}i$. Thus, for $x > 0$, $-\sqrt{-x} = -\sqrt{x}i$.

152 COMPLEX NUMBERS

EXAMPLES

$$\sqrt{-4} = \sqrt{4}i = 2i$$
$$-\sqrt{-25} = -\sqrt{25}i = -5i$$
$$-\sqrt{-18} = -\sqrt{18}i = -\sqrt{(9)(2)}i = -3\sqrt{2}i$$
$$\sqrt{-9}\sqrt{-25} = (\sqrt{9}i)(\sqrt{25}i) = (3i)(5i) = 15i^2 = -15$$
$$\sqrt{(-9)(-25)} = \sqrt{225} = 15$$
(Note that $\sqrt{-9}\sqrt{-25} \neq \sqrt{(-9)(-25)}$, since $\sqrt{-9}\sqrt{-25} = -15$, and $\sqrt{(-9)(-25)} = \sqrt{225} = 15$.)
$$\sqrt{-4} + \sqrt{-25} = 2i + 5i = 7i$$
$$\sqrt{-72} = \sqrt{72}i = \sqrt{(36)(2)}i = 6\sqrt{2}i$$
$$\sqrt{3}\sqrt{-2} = \sqrt{3}(\sqrt{2}i) = \sqrt{6}i$$

EXERCISES 7.5

Express each of the following in the form $a + bi$. Be sure all radicals are in simplest form.

1. $\sqrt{-5}$
2. $\sqrt{-3}$
3. $-\sqrt{-3}$
4. $-\sqrt{-5}$
5. $\sqrt{-8}$
6. $\sqrt{-7}\sqrt{-5}$
7. $\sqrt{-27}$
8. $\sqrt{-9}\sqrt{-4}$
9. $-2\sqrt{-6}$
10. $\sqrt{-7}\sqrt{-3}$
11. $(-\sqrt{5})(\sqrt{-7})$
12. $\sqrt{-3}\sqrt{-7}$
13. $-\sqrt{-32}$
14. $-\sqrt{-8}$
15. $(-\sqrt{3})(-\sqrt{-7})$
16. $\sqrt{-75} + \sqrt{-48}$
17. $\sqrt{-7} + \sqrt{-28} - \sqrt{-63}$
18. $\sqrt{-27} + \sqrt{-12}$
19. $\sqrt{-8} + \sqrt{-18}$
20. $\sqrt{-50} + \sqrt{-32}$

CHAPTER 7 REVIEW

1. Determine the real numbers, a and b, that satisfy the following.
    (a) $(1, a) = (b, -7)$
    (b) $a + 3i = -3 + bi$
    (c) $a + bi = 5i$
    (d) $3 - bi = a + 4i$
    (e) $2 - bi = a + 5i$
    (f) $5 + bi = a - 7i$

Given the four complex numbers, $z_1 = 2 + 3i$; $z_2 = -6$; $z_3 = -2i$; $z_4 = 3 - 4i$, find the following (Exercises 2-9).

2. $z_1 + z_2$
3. $z_1 - z_4$
4. $z_1 + z_4$
5. $z_1 \cdot z_4$
6. The conjugate of z_1
7. The conjugate of z_3
8. $z_3 - z_2$
9. $z_1 + z_2 + z_4$
10. What is the additive inverse of the complex number $3 - 4i$?
11. Give the real and imaginary parts of the following complex numbers.
    (a) $3 - 4i$
    (b) $2 + 5i$
    (c) 3
    (d) $-\sqrt{5}i$

Perform the indicated operations (Exercises 12–19).

12. $(3 + 2i) + (-3 - 4i)$

13. $(-2 - 3i) - (1 - i)$

14. $(2 + 3i)^2$

15. $(3 - 4i)(2 - i)$

16. $\dfrac{3 + i}{3 - i}$

17. $\dfrac{2 + i}{3 + 2i}$

18. $\dfrac{3 - 4i}{1 - 3i}$

19. $\dfrac{\sqrt{2} + 3i}{\sqrt{2} - 3i}$

20. Express the following in the form $a\sqrt{x}$ or $a\sqrt{x}i$, where a and $\sqrt{x}$ are real.
    (a) $\sqrt{-5}$
    (b) $\sqrt{-9}$
    (c) $\sqrt{2}\sqrt{-6}$
    (d) $\sqrt{-16}$
    (e) $\sqrt{-4}\sqrt{-9}$
    (f) $(-\sqrt{6})(-\sqrt{3})$

CHAPTER 8

Quadratic Functions

8.1 ◆ Definitions

We defined a linear function as a function whose values are given by a linear equation. Thus
$$\{(x, y) \mid y = ax + b\}$$
is a linear function.

We now define a **quadratic function** as a function whose values are given by a quadratic polynomial:
$$\{(x, y) \mid y = ax^2 + bx + c\}$$
where a, b, and c are real numbers and $a \neq 0$.

EXAMPLES

    I. $\{(x, y) \mid y = 2x^2 + 3x + 1\}$
   II. $\{(x, y) \mid y = 2x^2 - 1\}$
  III. $\{(x, y) \mid y = x^2\}$
  IV. $\{(x, y) \mid y = -x^2 - x - 1\}$

In function I, $a = 2$, $b = 3$, and $c = 1$. In II, $a = 2$, $b = 0$, and $c = -1$. In III, $a = 1$, $b = 0$, and $c = 0$. In IV, $a = -1$, $b = -1$, and $c = -1$.

EXERCISE 8.1

Identify the following as linear or quadratic functions.

1. $\{(x, y) \mid y = 3x^2\}$
2. $\{(x, y) \mid y + x = 3\}$
3. $\{(x, y) \mid y = 2x^2 - x + 1\}$
4. $\{(x, y) \mid y = 2x + 3\}$
5. $\{(x, y) \mid y = x^2 - 3x + 6\}$
6. $\{(x, y) \mid y + 3x + 4 = 0\}$
7. $\{(x, y) \mid y + x^2 = 0\}$
8. $\{(x, y) \mid y - 2 = x\}$
9. $\{(x, y) \mid y + x^2 = 2\}$
10. $\{(x, y) \mid y - 3x = x^2\}$

8.2 ◆ The Graph of a Quadratic Function

We now graph the quadratic function, beginning with a specific example:

$$\{(x, y) \mid y = x^2\}$$

We are seeking all the points (x, y) in the plane whose coordinates satisfy the equation $y = x^2$. Some of the ordered pairs that satisfy the equation $y = x^2$ are

$$(-3, 9), (-2, 4), (-1, 1), (0, 0), (1, 1), (2, 4), (3, 9)$$

Plotting the points with these ordered pairs as labels and drawing a smooth curve through these points, we obtain the graph in Figure 8.1. This curve is an example of a **parabola**.

It can be proved that the graph of any quadratic function

$$\{(x, y) \mid y = ax^2 + bx + c\} \qquad \text{(where } a \neq 0\text{)}$$

is a parabola and that the parabola opens upward if $a > 0$ and opens downward if $a < 0$. If the parabola opens upward, it is said to be **concave upward**. It then has a lowest point at which the function has its **minimum value**. If the parabola opens downward, it is said to be **concave downward** and has a highest point at which the function has its **maximum value**. The point at which a quadratic function has its maximum or minimum value is called the **vertex** of that parabola. From Figure 8.1 we see that the vertex of the parabola which is the graph of $y = x^2$ is the origin $(0, 0)$, that this parabola

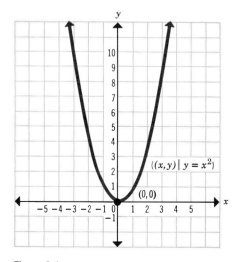

Figure 8.1

is concave up, and that the function specified by this equation has the minimum value 0.

EXAMPLE 1. Graph the function $\{(x, y) \mid y = x^2 + 2x + 1\}$.

SOLUTION. We have, by factoring, $y = (x + 1)^2$. Some of the ordered pairs which are elements of this function are

$$(-4, 9), (-3, 4), (-2, 1), (-1, 0), (0, 1), (1, 4), (2, 9)$$

Plotting these points and drawing a smooth curve connecting them, we have the graph in Figure 8.2. From this graph we see that the vertex of the parabola is $(-1, 0)$ and the minimum value of the function is 0.

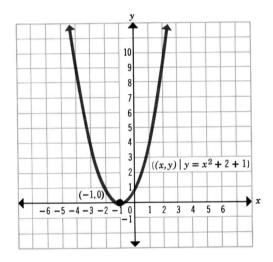

Figure 8.2

EXAMPLE 2. Graph the function $\{(x, y) \mid y = -x^2 + 1\}$.

SOLUTION. We have, as some of the ordered pairs that are elements of this function,

$$(-3, -8), (-2, -3), (-1, 0), (0, 1), (1, 0), (2, -3), (3, -8)$$

Plotting these points and drawing a smooth curve connecting them, we have the graph in Figure 8.3. The vertex of this parabola is $(0, 1)$. The maximum value of the function is 1.

8.3 COMPLETING THE SQUARE

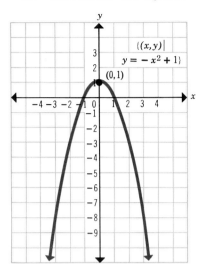

Figure 8.3

EXERCISES 8.2

Graph the quadratic functions specified by the equations below. Give the coordinates of the point where each function has a maximum or minimum value (Exercises 1–10).

1. $y = x^2$
2. $y = x^2 + 1$
3. $y = x^2 + 3$
4. $y = -x^2 + 1$
5. $y = -x^2 - 1$
6. $y = x^2 + 3x$
7. $y = 2x^2 - x$
8. $y = x^2 - x - 6$
9. $y = x^2 - 2x + 8$
10. $y = x^2 + 6x + 9$

11. Tell whether the graphs of the functions specified by the following equations open upward or downward.

   (a) $y = 3x^2 + 6x + 9$
   (b) $y = x^2 - 6$
   (c) $y = 6 - 3x^2$
   (d) $y = -2x^2 + 6$
   (e) $y = 2x^2 - 7x + 9$
   (f) $y = -2x^2 + 7$
   (g) $y = 7 + x^2$
   (h) $y = -4x^2$

8.3 ◆ Completing the Square

We now find a way to graph any quadratic function. We show that every equation of the form $y = ax^2 + bx + c$ is equivalent to an equation of the form $y = a(x - k)^2 + p$. In doing this, we will be able to find maximum and minimum values more easily.

We first look at some special cases. The first example is the function given by the equation $y = x^2 + 2x + 1$. Since $x^2 + 2x + 1 = (x + 1)^2$,

$$\{(x, y) \mid y = x^2 + 2x + 1\} = \{(x, y) \mid y = (x + 1)^2\}$$

That is, $y = x^2 + 2x + 1$ can be expressed in the equivalent form $y = a(x - k)^2 + p$, where $a = 1$, $k = -1$, and $p = 0$.

For a second example, we consider the function

$$\{(x, y) \mid y = 2x^2 + 4x + 2\}$$

Since $2x^2 + 4x + 2 = 2(x^2 + 2x + 1) = 2(x + 1)^2$, we see that

$$\{(x, y) \mid y = 2x^2 + 4x + 2\} = \{(x, y) \mid y = 2(x + 1)^2\}$$

and $y = 2x^2 + 4x + 2$ can be expressed in the equivalent form $y = a(x - k)^2 + p$, where $a = 2$, $k = -1$, and $p = 0$.

As our third example, we consider

$$\{(x, y) \mid y = x^2 - 6x + 10\}$$

We can determine the desired equivalent form of the equation $y = x^2 - 6x + 10$ by using the process called **completing the square**. In the case of $x^2 - 6x + 10$, the procedure is as follows.

$$\begin{aligned} x^2 - 6x + 10 &= (x^2 - 6x \quad) + 10 \\ &= (x^2 - 6x + 9) + 10 - 9 \\ &= (x - 3)^2 + 1 \end{aligned}$$

Notice that in the first step we simply separated the constant term, 10, from the terms involving the variable x. Then we determined what we must add to $x^2 - 6x$ to produce a perfect trinomial square, namely, 9 (the square of one-half the coefficient of x). Adding and subtracting 9 as indicated in the second step produces an equal expression, which gives us the desired form. Thus

$$\{(x, y) \mid y = x^2 - 6x + 10\} = \{(x, y) \mid y = (x - 3)^2 + 1\}$$

and we can write the equation $y = x^2 - 6x + 10$ in the form $y = a(x - k)^2 + p$, where $a = 1$, $k = 3$, and $p = 1$.

As a final specific example, we consider

$$\{(x, y) \mid y = 2x^2 + 4x - 5\}$$

The steps in obtaining the desired form equivalent to $y = 2x^2 + 4x - 5$ are

$$\begin{aligned} 2x^2 + 4x - 5 &= (2x^2 + 4x \quad) - 5 \\ &= 2(x^2 + 2x \quad) - 5 \\ &= 2(x^2 + 2x + 1) - 2 - 5 \\ &= 2(x + 1)^2 - 7 \end{aligned}$$

Notice that adding 1 to $x^2 + 2x$ within the parentheses was the same as adding 2 to the right member of the equation, so that we needed to subtract 2 to have the equation remain true. We conclude that

$$\{(x, y) \mid y = 2x^2 + 4x - 5\} = \{(x, y) \mid y = 2(x + 1)^2 - 7\}$$

The graphs of these four quadratic functions are shown in Figure 8.4.

Note that all three curves have similar shape. These graphs are examples of **parabolas**.

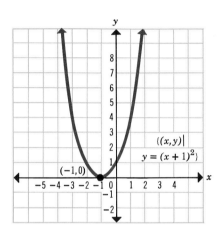
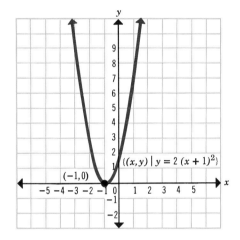
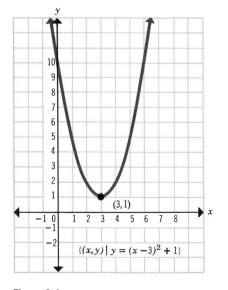
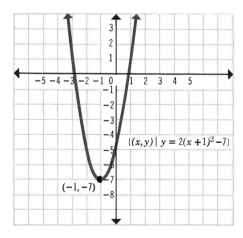

Figure 8.4

We now complete the square in the general quadratic function

$$\{(x, y) \mid y = ax^2 + bx + c\} \quad (a \neq 0)$$

$$\begin{aligned}
ax^2 + bx + c &= (ax^2 + bx \quad) + c \\
&= a\left(x^2 + \frac{b}{a}x \quad\right) + c \\
&= a\left(x^2 + \frac{b}{a}x + \frac{b^2}{4a^2}\right) - \frac{b^2}{4a} + c \\
&= a\left(x + \frac{b}{2a}\right)^2 + \frac{4ac - b^2}{4a}
\end{aligned}$$

The reader should justify each of these steps. As a result, we conclude that, in general, for $a \neq 0$,

$$\{(x, y) \mid y = ax^2 + bx + c\} = \{(x, y) \mid y = a(x - k)^2 + p\}$$

where

$$k = -\frac{b}{2a} \quad \text{and} \quad p = \frac{4ac - b^2}{4a}$$

We now analyze the graph of the general quadratic function by examining the equation

$$y = a\left(x + \frac{b}{2a}\right)^2 + \frac{4ac - b^2}{4a}$$

Since the square of a real number is never negative, $\left(x + \frac{b}{2a}\right)^2$ has zero for its least value, which occurs when x is $-\frac{b}{2a}$. For this value of x, y is $\frac{4ac - b^2}{4a}$.

Since $\left(x + \frac{b}{2a}\right)^2$ is either zero or positive, when $a > 0$, $a\left(x + \frac{b}{2a}\right)^2$ is zero or positive. Hence y will either be $\frac{4ac - b^2}{4a}$ or greater than $\frac{4ac - b^2}{4a}$. Therefore, the minimum value of the function is $\frac{4ac - b^2}{4a}$ when x is $-\frac{b}{2a}$.

When $a < 0$, $a\left(x + \frac{b}{2a}\right)^2$ is either zero or negative. Hence y will be either $\frac{4ac - b^2}{4a}$ or less than $\frac{4ac - b^2}{4a}$. Therefore the maximum of the function is $\frac{4ac - b^2}{4a}$ when x is $-\frac{b}{2a}$. The coordinates of the vertex of the parabola

8.3 COMPLETING THE SQUARE

which is the graph of the quadratic function $\{(x, y) \mid y = ax^2 + bx + c\}$ are the coordinates of the point where the function attains its maximum or minimum, that is,

$$(k, p) = \left(-\frac{b}{2a}, \frac{4ac - b^2}{4a}\right)$$

Once we know the coordinates of the vertex of the parabola, we can sketch the graph by finding the coordinates of several more points. Rather than try to remember the formula for the coordinates of the vertex, it is usually easiest to complete the square in each case.

EXAMPLE. Sketch the graph of $\{(x, y) \mid y = x^2 - 4x - 6\}$.

SOLUTION

$$\begin{aligned} x^2 - 4x - 6 &= (x^2 - 4x) - 6 \\ &= (x^2 - 4x + 4) - 4 - 6 \\ &= (x - 2)^2 - 10 \end{aligned}$$

The vertex of the parabola is $(2, -10)$. Since $a = 1 > 0$, the parabola opens up. The minimum of the function is -10. Two other points on the graph are $(1, -9)$ and $(3, -9)$. The graph is shown in Figure 8.5.

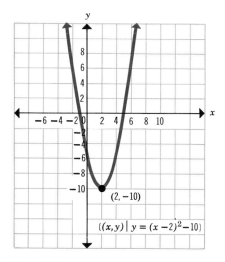

Figure 8.5

QUADRATIC FUNCTIONS

EXERCISES 8.3

Give the value of n for which the trinomial is a perfect square (Exercises 1–10).

1. $x^2 + 2x + n$
2. $x^2 - 12x + n$
3. $x^2 + 6x + n$
4. $x^2 - 20x + n$
5. $x^2 - 10x + n$
6. $x^2 + x + n$
7. $x^2 + 5x + n$
8. $x^2 + 16x + n$
9. $x^2 + 7x + n$
10. $x^2 + 8x + n$

Give the value which must be added to each member of the equation to make the expression in parentheses a perfect square (Exercises 11–20).

11. $y = (x^2 + 2x \quad)$
12. $y = (x^2 - 6x \quad)$
13. $y = 3(x^2 + 8x \quad)$
14. $y = 4(x^2 - 2x \quad)$
15. $y = -(x^2 - 4x \quad)$
16. $y = -2(x^2 + 6x \quad)$
17. $y = -16(x^2 + x \quad)$
18. $y = -4(x^2 - x \quad)$
19. $y = 3(x^2 - \frac{4}{3}x \quad)$
20. $y = 5(x^2 + \frac{3}{5}x \quad)$

Use the "completing the square" method to write an equivalent equation of the form $y = a(x - k)^2 + p$ (Exercises 21–30).

21. $y = x^2 + 12x + 36$
22. $y = 3x^2 + 6x + 3$
23. $y = x^2 + 6x + 5$
24. $y = x^2 + 10x + 11$
25. $y = -4x - x^2$
26. $y = -3x^2 + 30x - 73$
27. $y = 2x^2 - 12x + 19$
28. $y = -2x^2 - 8 - 8x$
29. $y = -33 - 16x - 2x^2$
30. $y = -5x^2 + 30x - 35$

Graph the functions specified by the equations below. In each case give the coordinates of the vertex (Exercises 31–36).

31. $y = x^2 + 10x + 11$
32. $y = -4x - x^2$
33. $y = -2x^2 - 8 + 8x$
34. $y = -5x^2 + 30x - 35$
35. $y = -33 - 16x - 2x^2$
36. $y = 2x^2 - 12x + 19$

8.4 ♦ Applications

Quadratic functions are often used to solve problems in which we construct a mathematical model of a physical situation. Consider the following.

EXAMPLE 1. An object is thrown into the air in such a way that its distance, d, in feet above the ground at the time t seconds after it was thrown is given by the equation $d = 64t - 16t^2$. Find the greatest height above the ground attained by the object.

SOLUTION. The equation given can be used to construct a quadratic function, namely,

$$\{(t, d) \mid d = 64t - 16t^2\}$$

If we complete the square, we have

$$\begin{aligned} 64t - 16t^2 &= -16t^2 + 64t \\ &= -16(t^2 - 4t) \\ &= -16(t^2 - 4t + 4) + 64 \\ &= -16(t - 2)^2 + 64 \end{aligned}$$

Some of the ordered pairs belonging to this function are

$$(-1, -80), (0, 0), (1, 48), (2, 64), (3, 48), (4, 0), (5, -80)$$

Plotting these points and using the information given by completing the square, we have the graph in Figure 8.6. The vertex of this parabola is $(2, 64)$, and the maximum value of the function is 64. The

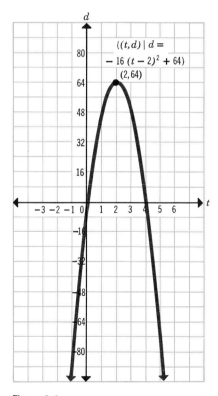

Figure 8.6

greatest height reached by the object is thus 64 feet, and this occurs 2 seconds after it has been thrown. Notice the function that describes the physical situation has a domain that is a subset of the domain of the quadratic function that we used as a model. The reason for this is only those points on the graph with d-coordinates that are positive or zero correspond to actual physical events.

EXAMPLE 2. A rectangular field along a riverbank is to be fenced out of a large parcel of property. No fence is needed along the river bank. What are the dimensions of the field of greatest area that can be fenced with 160 feet of fencing?

SOLUTION. We consider Figure 8.7.

The conditions of the problem require that

$$L + 2W = 160$$

We know that the area, A, of the field fenced in this way is given by

$$A = LW$$

The first equation relating L and W is equivalent to

$$L = 160 - 2W$$

Substituting into the second equation, we have

$$A = W(160 - 2W)$$
$$= 160W - 2W^2$$

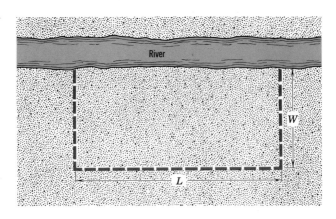

Figure 8.7

8.4 APPLICATIONS

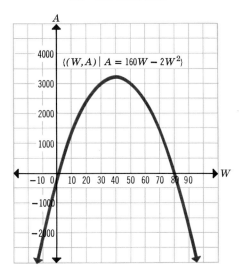

Figure 8.8

This equation gives us the quadratic function

$$\{(W, A) \mid A = 160W - 2W^2\}$$

Completing the square, we have

$$\begin{aligned}
160W - 2W^2 &= -2W^2 + 160W \\
&= -2(W^2 - 80W \quad) \\
&= -2(W^2 - 80W + 1600) + 3200 \\
&= -2(W - 40)^2 + 3200
\end{aligned}$$

The graph of this function (Figure 8.8) shows that the maximum area is 3200 square feet; it is obtained when the width is 40 feet. Hence the length is 80 feet.

EXERCISES 8.4

1. Find two numbers whose sum is 16 and whose product is a maximum. [*Hint.* Let x represent one number and $16 - x$ the other. The product, y, can be expressed as $y = x(16 - x)$.]
2. Find two numbers whose sum is 12 and whose product is as large as possible.
3. A boy wishes to fence the largest possible area for his dogs as a play area. He has 32 feet of fencing to use in making a rectangular area. Find the dimensions of the area.

QUADRATIC FUNCTIONS

4. Find the maximum area of a rectangle whose perimeter is 64 feet.
5. A man with 160 feet of fencing wishes to fence an area in the shape of a rectangle. What should be the dimensions of the area if the enclosed space is to be as large as possible?
6. Find two numbers whose sum is 20 such that the sum of their squares is a minimum.
7. Find the maximum area of a rectangle whose perimeter is 100 inches.
8. How close to the x-axis is the graph of $y = 2x^2 - 4x + 10$?
9. In a 110 volt circuit having a resistance of 11 ohms, the power, W, in watts delivered when a current of I amperes is flowing is given by $W = 110I - 11I^2$. Determine the maximum power that can be delivered by this circuit.
10. A real estate company estimates that the monthly profit, p, in dollars from a building s stories high is given by $p = -2s^2 + 88s$. What height of building would be considered the most profitable?

CHAPTER 8 REVIEW

Graph the quadratic functions given by the equations below. Give the coordinates of the vertex of each (Exercises 1–15).

1. $y = x^2 - 3$
2. $y = -x^2 + 1$
3. $y = x^2 - 4x + 4$
4. $y = x^2 - 4x + 1$
5. $y = -x^2 - 6x$
6. $y = -2x^2 - 4x$
7. $y = 8x - 2x^2$
8. $y = 3x^2 - 12x + 11$
9. $y = x^2 + x$
10. $y = 2x^2 + 3x + 4$
11. $2y = 4x^2 - 8x$
12. $2y = 2x^2 + x$
13. $3y = 12x - 3x^2$
14. $3y = 6x^2 + 6x + 10$
15. $4y = x^2 - x + 1$

16. What is the smallest product that can be produced by multiplying two consecutive odd integers?
17. A rectangular garden is to be fenced using 80 feet of fencing. What is the largest area it can have?
18. A small manufacturer finds that if he produces x intercoms, he will make a profit of $80 - 4x$ dollars on each. How many intercoms should he produce to get the largest profit?
19. If a projectile is fired straight up, its height above the ground is s feet after t seconds have elapsed. Experiments show that s and t are related by the equation $s = 80t - 16t^2$. When will the projectile hit the ground? How high will it go?
20. A rectangle is inscribed within the region bounded by the y-axis, the graph of $y = 1$, and the graph of $y = x^2$ by drawing horizontal and vertical lines

from a point on the parabola to the left and top boundaries of the region (Figure 8.9). Find the point on the parabola that produces the rectangle with the longest perimeter.

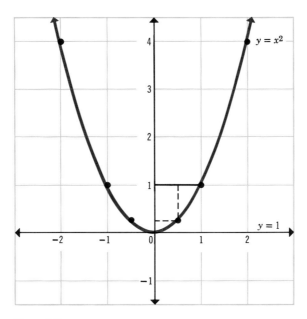

Figure 8.9

CHAPTER 9

Quadratic Equations and Inequalities

9.1 ♦ Quadratic Equations

Every equation of the form
$$ax^2 + bx + c = 0$$
where a, b, and c are real numbers and $a \neq 0$, is called a **quadratic equation**. In this book we shall restrict ourselves to the case in which a, b, and c are real numbers. The following equations are equivalent to quadratic equations:
$$3x^2 - 2x = 8$$
$$x^2 = 3$$
$$x^2 = 6 - 3x$$
$$7x^2 = 5x + 2$$
$$2x^2 = 15$$

In finding the solution set of a quadratic equation it is helpful to find an equivalent equation of the **standard form**: $ax^2 + bx + c = 0$. Thus, $3x^2 - 2x = 8$ is equivalent to the quadratic equation in standard form $3x^2 - 2x - 8 = 0$.

9.2 ♦ Solution Set of $x^2 = k$

Let us first consider quadratic equations for which $b = 0$. We see that we shall be considering equations that are equivalent to
$$x^2 = k$$
where k is a real number.

Since every nonzero real number has two square roots, we see that the solutions of the quadratic equation

9.2 SOLUTION SET OF $x^2 = k$

$$x^2 = k$$

$k \neq 0$, are the two square roots of k, namely, $\sqrt{k}$ and $-\sqrt{k}$. The solution set is $\{\sqrt{k}, -\sqrt{k}\}$.

If $k > 0$, the solutions of $x^2 = k$ are real numbers; if $k < 0$, the solutions are complex numbers.

If $k = 0$, we have

$$x^2 = 0$$

whose solution set is $\{0\}$.

EXAMPLE 1. Find the solution set of $x^2 = 9$ over the set R of real numbers.

SOLUTION. $x^2 = 9$ is equivalent to

$$x = \sqrt{9} \quad \text{or} \quad x = -\sqrt{9}$$
$$x = 3 \quad \text{or} \quad x = -3$$

The solution set is $\{-3, 3\}$.

EXAMPLE 2. Find the solution set of $4x^2 = 16$ over the set R of real numbers.

SOLUTION. $4x^2 = 16$ is equivalent to

$$x^2 = 4$$
$$x = \sqrt{4} \quad \text{or} \quad x = -\sqrt{4}$$
$$x = 2 \quad \text{or} \quad x = -2$$

The solution set is $\{-2, 2\}$.

EXAMPLE 3. Find the solution set of $5x^2 + 10 = 0$ over the set C of complex numbers.

SOLUTION. $5x^2 + 10 = 0$ is equivalent to

$$5x^2 = -10$$
$$x^2 = -2$$
$$x = \sqrt{-2} \quad \text{or} \quad x = -\sqrt{-2}$$
$$x = \sqrt{2}i \quad \text{or} \quad x = -\sqrt{2}i$$

The solution set is $\{-\sqrt{2}i, \sqrt{2}i\}$.

EXAMPLE 4. Find the solution set of $x^2 + 9 = 0$ over the set R of real numbers.

SOLUTION. $x^2 + 9 = 0$ is equivalent to

$$x^2 = -9$$

Since the square of a real number is a nonnegative real number, there are no real numbers which satisfy the equation $x^2 + 9 = 0$. The solution set over the set R of real numbers is $\emptyset$. (The solution set of the given equation over the set C of complex numbers is $\{-3i, 3i\}$.)

9.3 ◆ Solution by Factoring

In solving quadratic equations for which $b \neq 0$, we shall be considering equations of the form $ax^2 + bx + c = 0$ where a, b, and c are real numbers, $a \neq 0$. For some of these, the left member, $ax^2 + bx + c$, is readily factorable into the product of two first-degree polynomials over the integers.

In solving such equations over the set R of real numbers we shall make use of the following theorem about real numbers. **IF p AND q ARE REAL NUMBERS AND $pq = 0$, THEN $p = 0$ OR $q = 0$.** (Here the word "or" is used in the inclusive sense meaning one or the other or both.)

We shall consider two cases in which we find the solution set of quadratic equations by factoring.

CASE I. $b \neq 0$ and $c = 0$. Examples of such equations are

$$5x^2 + 3x = 0$$
$$x^2 + 2x = 0$$

When $c = 0$, $b \neq 0$, we are finding the solution set of an equation of the form

$$ax^2 + bx = 0$$

This equation is equivalent to an equation of the form

$$x^2 + kx = 0$$

where k is some real number. But $x^2 + kx = 0$ is equivalent to

$$x(x + k) = 0$$
$$x = 0 \quad \text{or} \quad x + k = 0$$
$$x = 0 \quad \text{or} \quad x = -k$$

The solution set is $\{0, -k\}$.

9.3 SOLUTION BY FACTORING

EXAMPLE 1. Find the solution set of $x^2 - 3x = 0$.

SOLUTION. $x^2 + 3x = 0$ is equivalent to

$$x(x - 3) = 0$$
$$x = 0 \quad \text{or} \quad x - 3 = 0$$
$$x = 0 \quad \text{or} \quad x = 3$$

The solution set is $\{0, 3\}$.

CASE II. $b \neq 0$, $c \neq 0$ and $ax^2 + bx + c$ is factorable into the product of two first degree binomials over the integers. Examples of such equations are

$$x^2 + 2x + 1 = 0$$
$$2x^2 + 5x - 3 = 0$$

In this case we are considering equations where the first member, $ax^2 + bx + c$, can be written as the product of two first degree binomials over the integers:

$$ax^2 + bx + c = (px + q)(rx + t) \quad p, q, r, \text{ and } t \text{ integers}$$

The given equation is then equivalent to

$$(px + q)(rx + t) = 0$$
$$px + q = 0 \quad \text{or} \quad rx + t = 0$$
$$x = \frac{-q}{p} \quad \text{or} \quad x = \frac{-t}{r}$$

The solution set is $\left\{\frac{-q}{p}, \frac{-t}{r}\right\}$.

EXAMPLE 2. Find the solution set of $x^2 - 3x - 4 = 0$ over R the set of real numbers.

SOLUTION. $x^2 - 3x - 4 = 0$ is equivalent to

$$(x - 4)(x + 1) = 0$$
$$x - 4 = 0 \quad \text{or} \quad x + 1 = 0$$
$$x = 4 \quad \text{or} \quad x = -1$$

The solution set is $\{-1, 4\}$.

EXAMPLE 3. Find the solution set of $3x^2 - 2x - 1 = 0$ over the set R of real numbers.

QUADRATIC EQUATIONS AND INEQUALITIES

SOLUTION. $3x^2 - 2x - 1 = 0$ is equivalent to

$$(3x + 1)(x - 1) = 0$$
$$3x + 1 = 0 \quad \text{or} \quad x - 1 = 0$$
$$x = -\tfrac{1}{3} \quad \text{or} \quad x = 1$$

The solution set is $\{-\tfrac{1}{3}, 1\}$.

EXAMPLE 4. Find the solution set of $2x^2 + x - 6 = 0$ over the set R of real numbers.

SOLUTION. $2x^2 + x - 6 = 0$ is equivalent to

$$(2x - 3)(x + 2) = 0$$
$$2x - 3 = 0 \quad \text{or} \quad x + 2 = 0$$
$$x = \tfrac{3}{2} \quad \text{or} \quad x = -2$$

The solution set is $\{-2, \tfrac{3}{2}\}$.

EXERCISE 9.1

Write each of the following equations in the form $ax^2 + bx + c = 0$, and determine the values of a, b, and c (Exercises 1–10).

1. $x^2 = 3x + 4$
2. $3x^2 - 2 = 4x$
3. $2x^2 = 6$
4. $x^2 + 2x = 0$
5. $x^2 - 3x = 9$
6. $7x^2 = 14$
7. $7 - x^2 - 5x = 0$
8. $1 + 3x = 2x^2$
9. $x^2 - x = 5$
10. $12x^2 + 8x = 3$

Find the solution sets of the following equations over the set C of complex numbers (Exercises 11–50).

11. $x^2 - 25 = 0$
12. $x^2 + 2 = 5$
13. $2x^2 = 1$
14. $3x^2 - 1 = 14$
15. $5x^2 - 1 = 4$
16. $x^2 + 3 = 28$
17. $x^2 - 36 = 0$
18. $x^2 + 2 = 0$
19. $x^2 - 3x = 0$
20. $x^2 = 5x$
21. $x^2 + 7x = 0$
22. $9x^2 - 100 = 0$
23. $\dfrac{x^2}{4} = 3$
24. $\dfrac{x^2}{5} - 3 = 0$
25. $3x^2 - 27 = 0$
26. $4x^2 - 5x = 0$
27. $5x = 15x^2$
28. $20x^2 - 15 = 0$
29. $2x^2 + 7 = 0$
30. $x^2 + 7x + 10 = 0$

31. $x^2 - 3x - 10 = 0$
32. $2x^2 + 7x + 3 = 0$
33. $4x^2 - 4x + 1 = 0$
34. $x^2 + 2x = 8$
35. $2x^2 - 5x + 2 = 0$
36. $2x^2 + x - 1 = 0$
37. $x^2 - x = 20$
38. $x^2 - 13x + 42 = 0$
39. $x^2 + 7x - 8 = 0$
40. $15x^2 + 4x - 4 = 0$
41. $2x^2 - 5x - 3 = 0$
42. $7x^2 + 17x - 12 = 0$
43. $x^2 - x = 2$
44. $6x^2 - x - 35 = 0$
45. $x^2 + 4x = 12$
46. $8x^2 - 6x = 9$
47. $6x^2 + 7x - 5 = 0$
48. $12x^2 - x - 1 = 0$
49. $24x^2 - 39x + 15 = 0$
50. $x^2 - ax - 12a^2 = 0$

9.4 • The Quadratic Formula

In some cases it is difficult or even impossible to factor the left member of $ax^2 + bx + c = 0$ into the product of two first degree binomials over the integers. The most useful method of solving such a quadratic equation is by the use of the quadratic formula. We shall obtain this formula by completing the square as we did in finding the vertex of the parabola which was the graph of a quadratic function. We want to find the solution set of

$$ax^2 + bx + c = 0$$

where $a \neq 0$. This equation is equivalent to:

$$ax^2 + bx = -c$$

$$x^2 + \frac{b}{a}x = -\frac{c}{a}$$

$$x^2 + \frac{b}{a}x + \frac{b^2}{4a^2} = \frac{b^2}{4a^2} - \frac{c}{a}$$

$$\left(x + \frac{b}{2a}\right)^2 = \frac{b^2 - 4ac}{4a^2}$$

$$x + \frac{b}{2a} = \frac{\sqrt{b^2 - 4ac}}{2a} \quad \text{or} \quad x + \frac{b}{2a} = -\frac{\sqrt{b^2 - 4ac}}{2a}$$

$$x = -\frac{b}{2a} + \frac{\sqrt{b^2 - 4ac}}{2a} \quad \text{or} \quad x = -\frac{b}{2a} - \frac{\sqrt{b^2 - 4ac}}{2a}$$

$$x = \frac{-b + \sqrt{b^2 - 4ac}}{2a} \quad \text{or} \quad x = \frac{-b - \sqrt{b^2 - 4ac}}{2a}$$

The solution set is thus

$$\left\{\frac{-b + \sqrt{b^2 - 4ac}}{2a}, \frac{-b - \sqrt{b^2 - 4ac}}{2a}\right\}$$

We may write the two solutions of $ax^2 + bx + c = 0$ in the abbreviated notation

$$\frac{-b \pm \sqrt{b^2 - 4ac}}{2a}$$

where the symbol "$\pm$" is read as "plus or minus." The equation

$$x = \frac{-b \pm \sqrt{b^2 - 4ac}}{2a}$$

which is equivalent to $ax^2 + bx + c = 0$, is called the **quadratic formula**. In the quadratic formula, the number $b^2 - 4ac$ is called the **discriminant** of the equation $ax^2 + bx + c = 0$ and affects the solution set of this equation in one of the following ways.

1. If $b^2 - 4ac = 0$, then there is only one real number solution.
2. If $b^2 - 4ac > 0$, then there are two distinct real number solutions.
3. If $b^2 - 4ac < 0$, then there are two complex number solutions which are conjugates.

The following examples illustrate the procedure for using the quadratic formula.

EXAMPLE 1. Find the solution set of $x^2 + x + 1 = 0$ over the set C of complex numbers.

SOLUTION. We have an equation of the form $ax^2 + bx + c = 0$ in which $a = 1$, $b = 1$, and $c = 1$. This equation is then equivalent to

$$x = \frac{-1 \pm \sqrt{1 - 4}}{2}$$

$$= \frac{-1 \pm \sqrt{-3}}{2}$$

$$= \frac{-1 \pm \sqrt{3}i}{2}$$

$$= -\tfrac{1}{2} \pm \tfrac{1}{2}\sqrt{3}i$$

The solution set is

$$\{-\tfrac{1}{2} + \tfrac{1}{2}\sqrt{3}i, \; -\tfrac{1}{2} - \tfrac{1}{2}\sqrt{3}i\}$$

EXAMPLE 2. Find the solution set of $9x^2 - 30x + 25 = 0$ over the set C of complex numbers.

9.4 THE QUADRATIC FORMULA

SOLUTION. We see that we have an equation in the desired form in which $a = 9$, $b = -30$, and $c = 25$. This equation is then equivalent to

$$x = \frac{-(-30) \pm \sqrt{(-30)^2 - 4(9)(25)}}{2(9)}$$

$$x = \frac{30 \pm \sqrt{900 - 900}}{18}$$

$$x = \frac{30}{18}$$

$$x = \frac{5}{3}$$

The solution set is $\{\frac{5}{3}\}$.

EXAMPLE. 3. Find the solution set of $2x^2 + 9x = 4$ over the set C of complex numbers.

SOLUTION. We begin by writing the equation in standard form:

$$2x^2 + 9x - 4 = 0$$

and note that $a = 2$, $b = 9$, and $c = -4$. This equation is then equivalent to

$$x = \frac{-9 \pm \sqrt{9^2 - 4(2)(-4)}}{2(2)}$$

$$x = \frac{-9 \pm \sqrt{81 + 32}}{4}$$

$$x = \frac{-9 \pm \sqrt{113}}{4}$$

The solution set is

$$\left\{\frac{-9 + \sqrt{113}}{4}, \frac{-9 - \sqrt{113}}{4}\right\}$$

EXERCISE 9.2

Find the solution sets over the set C of complex numbers using the quadratic formula (Exercises 1–30).

1. $x^2 - 2x - 7 = 0$
2. $x^2 - 2x - 12 = 0$
3. $x^2 + 4x - 28 = 0$
4. $x^2 - 2x - 8 = 0$
5. $2x^2 - 6x - 9 = 0$
6. $x^2 + 7x - 8 = 0$
7. $9x^2 - 30x + 23 = 0$
8. $9x^2 + 6x = 1$
9. $x^2 - x + 1 = 0$
10. $3x^2 - 3x + 1 = 0$

11. $x^2 + 4 = 0$
12. $x^2 - x + 2 = 0$
13. $3x^2 + 3x + 1 = 0$
14. $2x^2 + 5x + 3 = 0$
15. $x^2 + 3x = 9$
16. $2x^2 - 5x + 12 = 0$
17. $8x^2 + x - 9 = 0$
18. $2x^2 - x - 1 = 0$
19. $2x^2 - 2x + 1 = 0$
20. $x^2 + 3x + 5 = 0$
21. $9x^2 - 4x - 5 = 0$
22. $7x^2 = 9 - 5x$
23. $16x^2 - 4x + \frac{1}{4} = 0$
24. $5x^2 + 2x - 1 = 0$
25. $3x^2 = 2 - x$
26. $\frac{x^2 - 3}{2} = 1 - \frac{x}{4}$
27. $\frac{x^2}{3} = \frac{1}{2}x + \frac{3}{2}$
28. $x^2 - mx - 2m^2 = 0$
29. $3x^2 - kx + 4k^2 = 0$
30. $2x^2 - 7kx + 5k^2 = 0$

Use the discriminant to determine the character of the roots of each of the following equations (Exercises 31–40).

31. $x^2 + x + 1 = 0$
32. $x^2 + 6x + 9 = 0$
33. $2x^2 = 2x - 3$
34. $x^2 - 12x + 36 = 0$
35. $7x^2 = 9 - 5x$
36. $3x^2 = 5x + 2$
37. $x^2 + 5x - 6 = 0$
38. $2x^2 - 3x = 2$
39. $9x^2 - 6x + 1 = 0$
40. $2x^2 - 2x + 3 = 0$

Find the value of k that will make each of the following equations have only one number in their solution sets (Exercises 41–46). (*Hint*. Set $b^2 - 4ac = 0$.)

41. $x^2 - kx + 4 = 0$
42. $16x^2 + 16kx + 25 = 0$
43. $2x^2 - kx + 3 = 0$
44. $3x^2 - kx + 4 = 0$
45. $kx^2 - x + 3 = 0$
46. $kx^2 + x + 3 = 0$

9.5 • Properties of Roots of Quadratic Equations

In the preceding section, we saw that the two solutions, r_1 and r_2, called **roots** of the quadratic equation

$$ax^2 + bx + c = 0$$

are given by

$$r_1 = \frac{-b + \sqrt{b^2 - 4ac}}{2a} \quad \text{and} \quad r_2 = \frac{-b - \sqrt{b^2 - 4ac}}{2a}$$

The sum of the two roots is

$$r_1 + r_2 = \frac{-b + \sqrt{b^2 - 4ac}}{2a} + \frac{-b - \sqrt{b^2 - 4ac}}{2a}$$

9.5 PROPERTIES OF ROOTS OF QUADRATIC EQUATIONS

$$= \frac{-b + \sqrt{b^2 - 4ac} - b - \sqrt{b^2 - 4ac}}{2a}$$

$$= \frac{-2b}{2a}$$

$$= -\frac{b}{a}$$

The product of the two roots is

$$r_1 r_2 = \frac{-b + \sqrt{b^2 - 4ac}}{2a} \cdot \frac{-b - \sqrt{b^2 - 4ac}}{2a}$$

$$= \frac{(-b)^2 - (\sqrt{b^2 - 4ac})^2}{4a^2}$$

$$= \frac{b^2 - b^2 + 4ac}{4a^2}$$

$$= \frac{4ac}{4a^2}$$

$$= \frac{c}{a}$$

We see from these two results that the sum and product of the roots of the quadratic equation

$$ax^2 + bx + c = 0$$

depend on the values of a, b, and c—the coefficients of the polynomial which is the left member.

EXAMPLE 1. Find the sum and product of the roots of $3x^2 + 6x - 4 = 0$.

SOLUTION. We know that the sum of the roots of $ax^2 + bx + c = 0$ is $-\frac{b}{a}$. Here, $a = 3$ and $b = 6$. Thus the sum of the roots is $-\frac{6}{3} = -2$. Similarly, the product of the roots is $\frac{c}{a} = \frac{-4}{3} = -\frac{4}{3}$.

EXAMPLE 2. Write a quadratic equation with integer coefficients whose solution set is $\{\frac{3}{4}, \frac{1}{5}\}$.

SOLUTION. The sum of the roots is

$$\frac{3}{4} + \frac{1}{5} = \frac{19}{20} = -\frac{b}{a}$$

The product of the roots is

$$\frac{3}{4} \cdot \frac{1}{5} = \frac{3}{20} = \frac{c}{a}$$

The equation $ax^2 + bx + c = 0$ is equivalent to

$$x^2 + \frac{b}{a}x + \frac{c}{a} = 0$$

(since $a \neq 0$), so we know that

$$x^2 - \tfrac{19}{20}x + \tfrac{3}{20} = 0$$

is a quadratic equation whose solution set is $\{\tfrac{3}{4}, \tfrac{1}{5}\}$.
Since this equation is equivalent to

$$20x^2 - 19x + 3 = 0$$

we have found a quadratic equation with integer coefficients, which has the given set for its solution set.

EXERCISES 9.3

Give the sum and the product of the roots of the following quadratic equations (Exercises 1–10).

1. $3x^2 - 6x + 9 = 0$
2. $x^2 + 2x - 3 = 0$
3. $x^2 - 6x + 1 = 0$
4. $x^2 - 3x - 4 = 0$
5. $x^2 + x + 8 = 0$
6. $4x^2 - 3x + 1 = 0$
7. $3x^2 + 5x + 1 = 0$
8. $7x^2 + 5x - 1 = 0$
9. $5x^2 - 7x - 1 = 0$
10. $3x^2 + 5x + 3 = 0$

Write a quadratic equation in x whose solution set is given below (Exercises 11–20).

11. $\{2, 5\}$
12. $\{3, -2\}$
13. $\{\tfrac{1}{2}, \tfrac{1}{3}\}$
14. $\{\tfrac{3}{5}, \tfrac{2}{3}\}$
15. $\{-3, \tfrac{1}{2}\}$
16. $\{-\tfrac{1}{2}, -\tfrac{1}{4}\}$
17. $\{\tfrac{2}{3}, -\tfrac{3}{4}\}$
18. $\{\sqrt{5}, 2\sqrt{5}\}$
19. $\{-\sqrt{2}, \sqrt{2}\}$
20. $\{1 + \sqrt{2}, 1 - \sqrt{2}\}$

9.6 ◆ Fractional Equations

Some equations involving rational expressions may be solved by solving equivalent quadratic equations. In this case, care must be taken to be certain that the new equations are truly equivalent to the given ones.

EXAMPLE 1. Find the solution set of

$$\frac{x^2 + 12}{x - 3} = \frac{7x}{x - 3}$$

SOLUTION. In examining this equation, we see that 3 is not a permissible replacement for x, since $x - 3$ is 0 when x is replaced by 3. In a case such as this, the denominators of the left and right members of the equation vanish. We can find an equivalent equation by multiplying each member by $x - 3$ to obtain a new equation which contains no fractions. This procedure produces an equivalent equation *only* if we state that $x \neq 3$—that is, that $x - 3$ cannot be 0. We have

$$\frac{x^2 + 12}{x - 3} = \frac{7x}{x - 3}$$

$$(x - 3)\frac{x^2 + 12}{x - 3} = (x - 3)\frac{7x}{x - 3} \quad \text{and} \quad x \neq 3$$

$$x^2 + 12 = 7x \quad \text{and} \quad x \neq 3$$

$$x^2 - 7x + 12 = 0 \quad \text{and} \quad x \neq 3$$

$$(x - 4)(x - 3) = 0 \quad \text{and} \quad x \neq 3$$

$$x - 4 = 0$$

The solution set is $\{4\}$. Note that the solution set of the equation $(x - 4)(x - 3) = 0$ is $\{3, 4\}$, but we want the solution set of the open sentence which includes the restriction $x \neq 3$.

EXAMPLE 2. Find the solution set of

$$\frac{x}{x + 1} = \frac{-x}{x - 1}$$

SOLUTION. In examining this equation, we see that -1 and 1 are not permissible replacements for x, since these numbers would make the denominators of the members of this equation vanish. Again we find an equivalent equation whose solution set is easy to obtain. To do this we multiply each member by $(x + 1)(x - 1)$ to obtain a new equation which contains no fractions. This procedure produces an equivalent open sentence *only* if we also state that $x \neq 1$ and $x \neq -1$. We have

$$\frac{x}{x+1} = \frac{-x}{x-1}$$

$$(x+1)(x-1)\frac{x}{x+1} = (x+1)(x-1)\frac{-x}{x-1}, x \neq 1, \text{ and } x \neq -1$$

$$(x-1)x = (x+1)(-x), x \neq 1, \text{ and } x \neq -1$$

$$x^2 - x = -x^2 - x, x \neq 1, \text{ and } x \neq -1$$

$$2x^2 = 0, x \neq 1, \text{ and } x \neq -1$$

$$x^2 = 0, x \neq 1, \text{ and } x \neq -1$$

The solution set is $\{0\}$.

EXERCISES 9.4

For each equation below, tell what replacements for x are not permissible (Exercises 1–6).

1. $\dfrac{x}{3} + \dfrac{1}{x-2} = 0$

2. $\dfrac{x+1}{x-2} = \dfrac{x-2}{x+1}$

3. $\dfrac{2x-9}{x+4} = \dfrac{3x}{x-2}$

4. $\dfrac{2x+3}{x-7} = \dfrac{3}{x}$

5. $\dfrac{5}{x} = \dfrac{3}{x+2} + \dfrac{1}{3}$

6. $\dfrac{x+2}{x-2} = \dfrac{3}{x-1}$

Find the solution sets over R the set of real numbers (Exercises 7–18).

7. $\dfrac{3}{x+1} = \dfrac{x-2}{x+1}$

8. $\dfrac{5}{x} = \dfrac{x-3}{x+4}$

9. $\dfrac{x}{x-1} - \dfrac{1-x}{2} = \dfrac{2}{x-1}$

10. $\dfrac{2}{x-3} = \dfrac{3}{x+2} + \dfrac{4}{7}$

11. $\dfrac{x-5}{x+4} = \dfrac{x-1}{x-4}$

12. $\dfrac{7}{3x+1} = \dfrac{3}{2x-1}$

13. $\dfrac{3x-2}{2x-3} + \dfrac{2}{x-1} = 6$

14. $\dfrac{2}{4-x^2} - \dfrac{x}{4-x^2} = \dfrac{3}{x+2}$

15. $\dfrac{9}{x+1} - \dfrac{2x+3}{x-2} = \dfrac{7x-2x^2}{x^2-x-2}$

16. $\dfrac{3x-2}{2x+3} = \dfrac{9x-5}{6x+1}$

17. $\dfrac{7x}{x-3} - \dfrac{12x^2+12}{x^2-2x-3} = \dfrac{5x}{x+1}$

18. $\dfrac{x}{2x+1} - \dfrac{x+1}{2x-4} = \dfrac{2x-3}{2x^2-3x-2}$

9.7 ◆ Equations Containing Radicals

Many equations containing radicals can be solved by using quadratic equations. In this case we do not necessarily obtain a quadratic equation that is equivalent to the given equation, but one whose solution set contains all of the solutions of the given equation. It can be shown that IF EACH MEMBER OF AN EQUATION IS RAISED TO THE SAME POWER, THEN THE SOLUTION SET OF THE ORIGINAL EQUATION IS A SUBSET OF THE SOLUTION SET OF THE RESULTING EQUATION. Thus, if we square both members of an equation, the resulting equation is not necessarily equivalent to the given one and such a transformation is not an elementary one. We will illustrate this by considering the equation

$$\sqrt{x+2} = x - 4$$

Squaring each member, we have

$$(\sqrt{x+2})^2 = (x-4)^2$$

which is equivalent to

$$x + 2 = x^2 - 8x + 16$$
$$x^2 - 9x + 14 = 0$$
$$(x-2)(x-7) = 0$$

The solution set of $x^2 - 9x + 14 = 0$ is $\{2, 7\}$. We know that this set includes all of the solutions of the given equation. Upon checking each solution, we see that

$$\sqrt{2+2} = 2 - 4$$

is a false statement, since $\sqrt{4}$ is 2 and not -2. Thus 2 is not a solution of the given equation. But

$$\sqrt{7+2} = 7 - 4$$

is a true statement, since $\sqrt{9}$ is 3. Thus 7 is a solution of the given equation. Since no numbers other than 2 and 7 could be solutions (because the solution set of the new equation includes all the solutions of the given one), we conclude that the solution set of $\sqrt{x+2} = x - 4$ is $\{7\}$.

EXAMPLE 1. Find the solution set of $x + 2 = \sqrt{2x + 7}$.

SOLUTION. Squaring both members, we have

$$(x + 2)^2 = (\sqrt{2x + 7})^2$$
$$x^2 + 4x + 4 = 2x + 7$$
$$x^2 + 2x - 3 = 0$$
$$(x + 3)(x - 1) = 0$$

The solution set of the new equation is $\{1, -3\}$. Checking, we find that 1 is a solution of the given equation, but -3 is not. Thus the solution set of the given equation is $\{1\}$.

EXAMPLE 2. Find the solution set of $\sqrt{x + 14} = 1 + \sqrt{2x + 5}$.

SOLUTION. Squaring both members, we have

$$(\sqrt{x + 14})^2 = (1 + \sqrt{2x + 5})^2$$
$$x + 14 = 1 + 2\sqrt{2x + 5} + 2x + 5$$
$$-x + 8 = 2\sqrt{2x + 5}$$

Again squaring both members, we have

$$x^2 - 16x + 64 = 4(2x + 5)$$
$$x^2 - 16x + 64 = 8x + 20$$
$$x^2 - 24x + 44 = 0$$
$$(x - 2)(x - 22) = 0$$

The solution set of the new equation is $\{2, 22\}$. Checking each of these solutions in the given equation, we see that its solution set is $\{2\}$.

EXERCISES 9.5

Find the solution sets over R the set of real numbers. Verify all solutions.

1. $\sqrt{x + 2} = \sqrt{2x - 3}$
2. $\sqrt{x + 4} = 5$
3. $\sqrt{x + 4} - 2 = 0$
4. $\sqrt{x^2 - 1} = x - 1$
5. $2\sqrt{x - 1} = 3$
6. $\sqrt{x^2 + 2} = 2 - x$
7. $\sqrt{4x + 3} = \sqrt{6x + 2}$
8. $\sqrt{2x - 5} = 1 + \sqrt{x - 3}$
9. $\sqrt{2 - 8x} = 2 - 2\sqrt{1 - 6x}$
10. $\sqrt{x + 2} + \sqrt{x - 1} = \sqrt{2x + 5}$

9.8 ◆ Equations with Quadratic Form

There are many equations that are not themselves quadratic, but that are equivalent to an equation having the form

$$aw^2 + bw + c = 0 \quad (a \neq 0)$$

9.8 EQUATIONS WITH QUADRATIC FORM

where w represents an expression in another variable. Some examples are

(a) $x - 10\sqrt{x} + 9 = 0$
(b) $x^4 + 2x^2 + 1 = 0$
(c) $8x^{-2} + 7x^{-1} - 1 = 0$

These equations are said to have **quadratic form,** since:

(a) $x - 10\sqrt{x} + 9 = 0$ is equivalent to $(\sqrt{x})^2 - 10(\sqrt{x}) + 9 = 0$
(b) $x^4 + 2x^2 + 1 = 0$ is equivalent to $(x^2)^2 + 2(x^2) + 1 = 0$
(c) $8x^{-2} + 7x^{-1} - 1 = 0$ is equivalent to $8(x^{-1})^2 + 7(x^{-1}) - 1 = 0$

We can solve equations having quadratic form by first finding the solutions of the equivalent quadratic equation in the new variable and then solving the resulting equations.

For example, in (a),

$$x - 10\sqrt{x} + 9 = 0$$

if we let $w = \sqrt{x}$, we have the equivalent equations

$$w^2 - 10w + 9 = 0$$
$$(w - 1)(w - 9) = 0$$

The solution set of the new equation in the variable w is $\{1, 9\}$. The given equation is then equivalent to

$$\sqrt{x} = 1 \quad \text{or} \quad \sqrt{x} = 9$$
$$x = 1 \quad \text{or} \quad x = 81$$

from which we obtain the solution set $\{1, 81\}$.

EXAMPLE 1. Find the solution set of $x^4 - 5x^2 + 4 = 0$.

SOLUTION. Let $w = x^2$. We have

$$w^2 - 5w + 4 = 0$$
$$(w - 1)(w - 4) = 0$$
$$w = 1 \quad \text{or} \quad w = 4$$

Then

$$x^2 = 1 \quad \text{or} \quad x^2 = 4$$
$$x = \pm 1 \quad \text{or} \quad x = \pm 2$$

The solution set is $\{-2, -1, 1, 2\}$.

EXAMPLE 2. Find the solution set of $x^{-2} + 3x^{-1} - 4 = 0$.

SOLUTION. We let $w = x^{-1}$. Then

$$w^2 + 3w - 4 = 0$$
$$(w + 4)(w - 1) = 0$$
$$w = -4 \quad \text{or} \quad w = 1$$

Then

$$x^{-1} = -4 \quad \text{or} \quad x^{-1} = 1$$
$$\frac{1}{x} = -4 \quad \text{or} \quad \frac{1}{x} = 1$$
$$x = -\frac{1}{4} \quad \text{or} \quad x = 1$$

The solution set is $\{-\frac{1}{4}, 1\}$.

EXAMPLE 3. Find the solution set of $4 + 3x^{1/4} - x^{1/2} = 0$.

SOLUTION. Let $w = x^{1/4}$. Then

$$4 + 3w - w^2 = 0$$
$$(4 - w)(1 + w) = 0$$
$$w = 4 \quad \text{or} \quad w = -1$$

Then $x^{1/4} = 4$ or $x^{1/4} = -1$. Since x must be a real number, because we are finding the solution set over R the set of real numbers, $x^{1/4} = -1$ is impossible and must be rejected. Why? Then

$$x^{1/4} = 4$$
$$(x^{1/4})^4 = 4^4$$
$$x = 256$$

The solution set is $\{256\}$.

EXERCISES 9.6

Find the solution sets over the set R of real numbers. Verify each solution.

1. $x^4 - 3x^2 + 2 = 0$
2. $x^4 - 4x^2 + 3 = 0$
3. $x^4 - 10x^2 + 9 = 0$
4. $x^4 - 9x^2 + 8 = 0$
5. $x^4 - 9x^2 + 18 = 0$
6. $x^4 = 10x^2 - 25$
7. $x - 17\sqrt{x} + 16 = 0$
8. $x - 5\sqrt{x} + 4 = 0$
9. $2x - 9\sqrt{x} + 4 = 0$
10. $6x + \sqrt{x} - 1 = 0$
11. $2x + 4 = 9\sqrt{x}$
12. $x - \sqrt{x} - 20 = 0$

13. $x^{-4} - 5x^{-2} + 4 = 0$
14. $x^{-2} - 2x^{-1} + 1 = 0$
15. $x^{-4} = 3x^{-2} + 4$
16. $2x^{-4} + 3 = 5x^{-2}$
17. $4x^{-4} + 4 - 17x^{-2} = 0$
18. $4x^{-4} - 16x^{-2} + 15 = 0$
19. $x^{1/2} - 2x^{1/4} = 8$
20. $5 + 3x^{1/4} = 8x^{1/2}$
21. $x^{3/2} + 7x^{3/4} - 8 = 0$
22. $x^{-2} - x^{-1} = 12$
23. $x^{2/3} - 35 = 2x^{1/3}$
24. $x^{-2} + 9x^{-1} - 10 = 0$
25. $(x - 5)^2 - 5(x - 5) + 6 = 0$
26. $(x^2 + 5x)^2 - 8(x^2 + 5x) = 84$
27. $\sqrt[4]{x - 2} + \sqrt{x - 2} = 2$
28. $4(x^2 - 1)^2 - 10(x^2 - 1) + 6 = 0$
29. $(2x^2 - 7)^2 - 12(2x^2 - 7) + 11 = 0$
30. $(3x^2 - 7x)^2 + 6(3x^2 - 7x) + 8 = 0$

9.9 ◆ Quadratic Inequalities

When we solved quadratic equations over the set R of real numbers, we used the fact that the product of two factors is zero if and only if one or both of the factors are zero. To solve quadratic inequalities over R we use the following property about real numbers.

The product of two real numbers is positive if and only if both numbers are positive or both numbers are negative; the product of two real numbers is negative if and only if one of the real numbers is positive and the other is negative.

EXAMPLE 1. Find the solution set of $x^2 - 2x - 3 > 0$ over the set R of real numbers.

SOLUTION. The inequality $x^2 - 2x - 3 > 0$ is equivalent to

$$(x + 1)(x - 3) > 0$$

Since the product of two real number factors is positive, we have two cases:

Case 1. $x + 1 > 0$ and $x - 3 > 0$

Case 2. $x + 1 < 0$ and $x - 3 < 0$

In Case 1 we have $x + 1 > 0$ and $x - 3 > 0$ which is equivalent to

$x > -1$ and $x > 3$ which is equivalent to
$x > 3$

In Case 2 we have $x + 1 < 0$ and $x - 3 < 0$ which is equivalent to

$x < -1$ and $x < 3$ which is equivalent to
$x < -1$

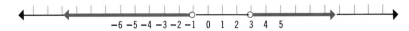

Figure 9.1

The given inequality is equivalent to

$$x > 3 \quad \text{or} \quad x < -1$$

and the solution set is

$$\{x \mid x < -1 \quad \text{or} \quad x > 3\}$$

We graph this solution set on the number line (see Figure 9.1).

EXAMPLE 2. Find the solution set of $x^2 - 4x + 3 < 0$.

SOLUTION. $x^2 - 4x + 3 < 0$ is equivalent to

$$(x-1)(x-3) < 0$$

Case 1. $x - 3 < 0$ and $x - 1 > 0$
Case 2. $x - 3 > 0$ and $x - 1 < 0$

Case 1. $x - 3 < 0$ and $x - 1 > 0$ is equivalent to

$$x < 3 \quad \text{and} \quad x > 1, \text{ which is equivalent to } 1 < x < 3$$

Case 2. $x - 3 > 0$ and $x - 1 < 0$ is equivalent to

$$x > 3 \quad \text{and} \quad x < 1$$

There are no real numbers satisfying this open sentence.
The solution set is

$$\{x \mid 1 < x < 3\}$$

The graph of this solution set is shown in Figure 9.2.

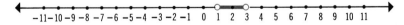

Figure 9.2

EXERCISES 9.7

Find the solution sets over R the set of real numbers and graph.

1. $x^2 > 9$ 2. $x^2 < 16$

CHAPTER 9 REVIEW

3. $x^2 + 6 > 10$
4. $x^2 - 3 < 6$
5. $x^2 \geq 25$
6. $x^2 \leq 36$
7. $x^2 - 5 \leq 4$
8. $x^2 + 12 \leq 13$
9. $(x - 2)(x + 3) > 0$
10. $(x + 1)(x + 2) < 0$
11. $(x - 3)(x + 2) \geq 0$
12. $(x + 4)(x + 5) < 0$
13. $(x - 1)(x + 3) > 0$
14. $x^2 + 3x > 0$
15. $x^2 - 3x + 2 > 0$
16. $x^2 - 8x + 15 > 0$
17. $x^2 - 5x + 6 \geq 0$
18. $x^2 - 5x + 4 < 0$
19. $2x^2 - 7x + 6 < 0$
20. $15x^2 - 7x - 2 < 0$

CHAPTER 9 REVIEW

Find the solution sets.

1. $x^2 = 16$
2. $x^2 + 9 = 0$
3. $x^2 - 13x + 30 = 0$
4. $x^2 + 10x + 16 = 0$
5. $4x^2 + x - 14 = 0$
6. $x^2 + 6x - 13 = 0$
7. $5x^2 + 22x + 8 = 0$
8. $3x^2 + 10x + 2 = 0$
9. $x^2 = 6x - 10$
10. $4x^2 - 4x = 5$
11. $\dfrac{2}{x - 2} = 2 - \dfrac{3}{x + 1}$
12. $\dfrac{x}{2} + \dfrac{2}{x} = 2x$
13. $\sqrt{x + 3} = 6$
14. $2\sqrt{x + 1} = 4\sqrt{x - 5}$
15. $\sqrt{x + 1} + \sqrt{x + 2} = \sqrt{2x + 3}$
16. $x^4 - 10x^2 + 25 = 0$
17. $4x^{-4} - 17x^{-2} + 4 = 0$
18. $(x - 3)(x + 3) > 0$
19. $x^2 - x - 6 < 0$
20. $x^2 + 10x + 21 \geq 0$

CHAPTER 10

Systems of Equations

10.1 ◆ Equations in Two Variables

An equation of the form

$$ax + by + c = 0$$

is called an **equation in two variables**. The **solution set** of such an equation is the set of all ordered pairs that satisfy the equation. Since this equation specifies a linear function, its graph is a straight line. Because of this, such an equation is often called a **linear equation.**

It is often necessary to consider pairs of such equations and to inquire whether or not the solution sets of the two equations contain ordered pairs in common. That is, we are interested in the set of all ordered pairs satisfying both equations. The pair of equations

$$a_1 x + b_1 y + c_1 = 0 \qquad (a_1 \text{ and } b_1 \text{ not both } 0)$$
$$a_2 x + b_2 y + c_2 = 0 \qquad (a_2 \text{ and } b_2 \text{ not both } 0)$$

is called a **system** of two linear equations in two variables.

We are interested in determining the members of the **intersection** of their solution sets, that is, the set of ordered pairs common to both the solution sets.

10.2 ◆ Independent, Inconsistent, and Dependent Systems of Equations

Let us consider the system

$$a_1 x + b_1 y + c_1 = 0 \qquad (a_1 \text{ and } b_1 \text{ not both } 0)$$
$$a_2 x + b_2 y + c_2 = 0 \qquad (a_2 \text{ and } b_2 \text{ not both } 0)$$

10.2 INDEPENDENT, INCONSISTENT, AND DEPENDENT SYSTEMS OF EQUATIONS

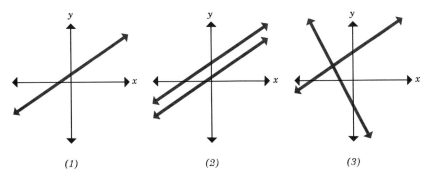

Figure 10.1

We want to find the set of all ordered pairs of real numbers each satisfying both equations.

In a geometric sense, because the graphs of both equations are straight lines, we are confronted with the three possibilities shown in Figure 10.1.

1. The graphs are the same line.
2. The graphs are parallel but distinct lines.
3. The graphs are intersecting lines.

These possibilities lead to the following conclusion: one and only one of the following is true for any pair of linear equations in two variables.

1. The solution sets of the equations are equal, and every ordered pair that satisfies one of the given equations satisfies the system.

2. The two solution sets have no ordered pairs in common, that is, the empty set is the solution set of the system.

3. The two solution sets have one and only one ordered pair in common, that is, the system has one and only one solution.

In Case 1 we say that the system is a **dependent** system; in Case 2 we say that a system is an **inconsistent** system; and in Case 3 we say that the system is an **independent** system. If a system is a dependent system, the left member of one of the equations can be obtained from the left member of the other through multiplication by a constant. Thus

$$2x + 3y + 6 = 0$$
$$6x + 9y + 18 = 0$$

is a dependent system because

$$6x + 9y + 18 = 3(2x + 3y + 6)$$

SYSTEMS OF EQUATIONS

EXERCISES 10.1

Graph the equations of each system. Tell whether each system is independent, inconsistent, or dependent.

1. $x + y = 3$
   $2x + 2y = 6$
2. $x + y = 7$
   $x - y = 3$
3. $x + 2y = 6$
   $x + 2y = 8$
4. $x + y = 3$
   $5x - y = 10$
5. $x + y = 0$
   $2x + 3y + 6 = 0$
6. $x = y$
   $x = 2y$
7. $3x + y = 6$
   $x + 2y = 8$
8. $x + 3y = 4$
   $x - 3y = 4$
9. $x + 2y = 8$
   $2y = x - 4$
10. $5x - y = 10$
    $3x - 5y = 15$

10.3 ♦ Solution by Graphing

A system of two linear equations has a graph that consists of two straight lines—the graphs of each of the equations of the system. If the two lines intersect, the system of two linear equations has one and only one ordered pair for its solution. If the two lines are parallel, the system of two linear equations has the empty set as its solution set. If the two lines are coincident, that is, the two equations have the same graph, the system of two linear equations has an infinite set of ordered pairs for its solution set.

We can arrive at the solution set of a system by graphing the equations. It must be noted that solutions arrived at graphically are, at best, approximate, and should be checked in each equation of the system. The procedure for finding the solutions of a system graphically is illustrated below.

EXAMPLE 1. Solve by graphing:

$$x + 2y = 4$$
$$3x - 2y = 4$$

SOLUTION. From the graph (Figure 10.2) we see that the two lines appear to intersect at $(2, 1)$. We check to see whether this pair satisfies each equation:

$$2 + 2(1) = 4 \qquad 3(2) - 2(1) = 4$$
$$4 = 4 \qquad\qquad 4 = 4$$

The solution set is $\{(2, 1)\}$.

10.3 SOLUTION BY GRAPHING

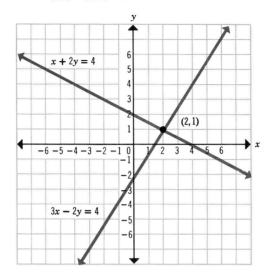

Figure 10.2

EXAMPLE 2. Solve by graphing:

$$x + y = 3$$
$$2x + 2y = 4$$

SOLUTION. The graph of this system (Figure 10.3) consists of two parallel lines. The solution set is the empty set, $\varnothing$.

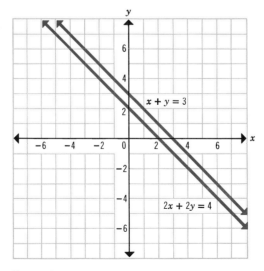

Figure 10.3

SYSTEMS OF EQUATIONS

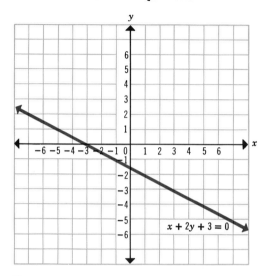

Figure 10.4

EXAMPLE 3. Solve graphically:

$$x + 2y + 3 = 0$$
$$2x + 4y + 6 = 0$$

SOLUTION. The two equations have the same graph (Figure 10.4). Thus the solution set of the system is an infinite set of ordered pairs. Some elements of the solution set are $(-3, 0)$, $(1, -2)$, and $(5, -4)$. Using set builder notation we denote the solution set by $\{(x, y) \mid x + 2y + 3 = 0\}$.

EXERCISES 10.2

Solve each system by graphing. Check the solutions you obtain (Exercises 1–15).

1. $x + y = 6$
   $x - y = 4$
2. $x + y = 6$
   $x - y = 2$
3. $3x + y = 6$
   $2x - y = 4$
4. $2x + y = 6$
   $x - 2y = -2$
5. $2x + y = 8$
   $4x - y = 4$

6. $x - y + 2 = 0$
   $2x + 4y - 26 = 0$
7. $2x - y = 4$
   $2x + y = 8$
8. $x - 2y = 3$
   $3x = 6y + 9$
9. $2x - 6y = 2$
   $2x + y = 9$
10. $y = 2x - 9$
    $y = -3x + 11$

11. $x - y + 5 = 0$
    $2x + y - 5 = 0$
12. $2x + y = 6$
    $3x + 4y = 4$
13. $2x - y - 8 = 0$
    $x - 3y - 4 = 0$
14. $3x - 2y = 7$
    $x + 2y = 5$
15. $x + 2y - 3 = 0$
    $4x + 8y - 12 = 0$

10.4 ♦ Analytic Solution

Because graphic results are, at best, approximations, solutions of systems of linear equations are usually found by analytical methods. Solving systems of equations analytically depends on the following theorem which we shall accept without proof.

> IF TWO EQUATIONS OF A SYSTEM ARE COMBINED, TERM BY TERM, THROUGH ADDITION OR SUBTRACTION, OR EACH IS MULTIPLIED BY SOME CONSTANT, NOT ZERO, AND THEN COMBINED, THE RESULT WILL BE A NEW LINEAR EQUATION WHOSE SOLUTION SET CONTAINS THE SOLUTION OR SOLUTIONS OF THE ORIGINAL SYSTEM.

We shall illustrate the procedure used to find solution sets of systems of linear equations in two variables in the following examples.

EXAMPLE 1. Find the solution set of the system

(1) $\qquad 7x + 3y = 4$
(2) $\qquad 14x + 9y = 7$

SOLUTION. We multiply each member of equation (1) by 2 and each member of equation (2) by 1 to obtain

(3) $\qquad 14x + 6y = 8$
(4) $\qquad 14x + 9y = 7$

Subtracting the left and right members of equation (3) from those of equation (4) we obtain

$$3y = -1$$

which is equivalent to

$$y = -\tfrac{1}{3}$$

Replacing y by $-\tfrac{1}{3}$ in equation (1), we have

$$7x + 3(-\tfrac{1}{3}) = 4$$
$$7x - 1 = 4$$
$$7x = 5$$
$$x = \tfrac{5}{7}$$

The solution set of the system is

$$\{(\tfrac{5}{7}, -\tfrac{1}{3})\}$$

CHECK

$$7(\tfrac{5}{7}) + 3(-\tfrac{1}{3}) = 4 \qquad 14(\tfrac{5}{7}) + 9(-\tfrac{1}{3}) = 7$$
$$5 + (-1) = 4 \qquad 10 + (-3) = 7$$
$$4 = 4 \qquad 7 = 7$$

EXAMPLE 2. Find the solution set of

(5) $\qquad x + 4y = 0$
(6) $\qquad 4x - y = 17$

SOLUTION. We multiply both members of equation (5) by 4 and both members of equation (6) by 1 to obtain.

(7) $\qquad 4x + 16y = 0$
(8) $\qquad 4x - y = 17$

We subtract the members of equation (8) from the members of equation (7) which gives

$$17y = -17$$
$$y = -1$$

Replacing y by -1 in $x + 4y = 0$ gives

$$x + 4(-1) = 0$$
$$x = 4$$

The solution set is $\{(4, -1)\}$.

CHECK

$$4 + 4(-1) = 0 \qquad 4(4) - (-1) = 17$$
$$4 - 4 = 0 \qquad 16 + 1 = 17$$

EXAMPLE 3. Find the solution set of

$$2x - 3y - 20 = 0$$
$$-3x + 5y + 11 = 0$$

SOLUTION. We multiply both members of the first equation by 5 and both members of the second equation by 3 to obtain

$$10x - 15y - 100 = 0$$
$$-9x + 15y + 33 = 0$$

Adding the members of these two equations, we obtain

$$x - 67 = 0$$
$$x = 67$$

10.4 ANALYTIC SOLUTION

Replacing x by 67 in $2x - 3y - 20 = 0$ gives

$$2(67) - 3y - 20 = 0$$
$$134 - 3y - 20 = 0$$
$$114 - 3y = 0$$
$$3y = 114$$
$$y = 38$$

The solution set is $\{(67, 38)\}$.

CHECK

$$2(67) - 3(38) - 20 = 0 \qquad -3(67) + 5(38) + 11 = 0$$
$$134 - 114 - 20 = 0 \qquad -201 + 190 + 11 = 0$$

EXERCISES 10.3

Find the solution sets of the following systems analytically.

1. $2x + y = 7$
   $3x - y = 8$
2. $x + y = 6$
   $x - y = 4$
3. $x + 3y = 11$
   $x - 5y = -13$
4. $9x + 8y = 3$
   $9x - 8y = -93$
5. $3x + y = 5$
   $6x - y = 6$
6. $5x + 4y = 22$
   $3x + y = 9$
7. $3x + 4y = 6$
   $x - 2y = -8$
8. $x + 3y = 9$
   $4x + 5y = 22$
9. $x + 2y = 4$
   $4x + 8y = 16$
10. $4x - 2y = 5$
    $4x + 7y = 4$
11. $4x - 7y = 10$
    $5x + 3y = 11$
12. $7x - y = 4$
    $-14x + 2y = -8$
13. $5x + 2y = 8$
    $7x - y = 15$
14. $4x - 5y = 26$
    $5x + 3y = 51$
15. $6x - 5y + 4 = 0$
    $x - 10y + 1 = 0$
16. $-2x + 3y = 10$
    $3x + 5y = -15$
17. $2x - 3y = 1$
    $3x - 4y = 7$
18. $4x - y = -10$
    $3x + 5y = 4$
19. $\dfrac{x}{2} - \dfrac{y}{5} = \dfrac{21}{10}$
    $\dfrac{x}{6} + \dfrac{y}{4} = -\dfrac{1}{4}$
20. $\dfrac{8}{5}x - y = 1$
    $\dfrac{x}{2} - \dfrac{y}{3} = \dfrac{29}{6}$

10.5 ◆ Determinants

A symbol such as the one in Figure 10.5 is called a **determinant**. The symbols a_1, a_2, b_1, and b_2 represent numbers and are called the **elements**

$$\begin{vmatrix} a_1 & b_1 \\ a_2 & b_2 \end{vmatrix}$$

Figure 10.5

of the determinant. Because this determinant has two rows and two columns, we call it a 2×2 (read: two by two), or **second order** determinant. We name the rows and columns as shown in Figure 10.6.

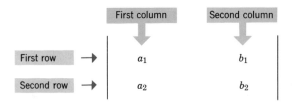

Figure 10.6

We also name the diagonals of a 2×2 determinant as shown in Figure 10.7.

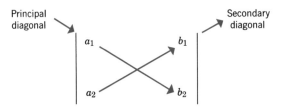

Figure 10.7

The determinant $\begin{vmatrix} a_1 & b_1 \\ a_2 & b_2 \end{vmatrix}$ names a number. This number is obtained by multiplying the elements on the two diagonals and adding the opposite of the secondary product to the principal product. This process can be shown schematically as shown below.

$$\begin{vmatrix} a_1 & b_1 \\ a_2 & b_2 \end{vmatrix} = a_1 b_2 - a_2 b_1$$

EXAMPLES

$$\begin{vmatrix} 3 & 4 \\ -2 & 1 \end{vmatrix} = (3)(1) - (-2)(4) = 3 + 8 = 11$$

$$\begin{vmatrix} 1 & -3 \\ -2 & -4 \end{vmatrix} = (1)(-4) - (-2)(-3) = -4 - 6 = -10$$

$$\begin{vmatrix} 0 & 3 \\ 1 & -2 \end{vmatrix} = (0)(-2) - (1)(3) = 0 - 3 = -3$$

EXERCISES 10.4

Find the number named by each of the following determinants.

1. $\begin{vmatrix} 1 & 2 \\ 3 & 2 \end{vmatrix}$

2. $\begin{vmatrix} 3 & 0 \\ -1 & 2 \end{vmatrix}$

3. $\begin{vmatrix} 1 & 0 \\ 0 & 1 \end{vmatrix}$

4. $\begin{vmatrix} 9 & -2 \\ 4 & -6 \end{vmatrix}$

5. $\begin{vmatrix} -1 & 6 \\ 0 & -2 \end{vmatrix}$

6. $\begin{vmatrix} 1 & -3 \\ -1 & 3 \end{vmatrix}$

7. $\begin{vmatrix} \frac{1}{4} & \frac{1}{2} \\ -\frac{1}{2} & \frac{3}{4} \end{vmatrix}$

8. $\begin{vmatrix} i & -i \\ -i & 2i \end{vmatrix}$

9. $\begin{vmatrix} 0 & 0 \\ 7 & 6 \end{vmatrix}$

10. $\begin{vmatrix} 5i & -3 \\ -2 & 2i \end{vmatrix}$

11. $\begin{vmatrix} 0 & -3 \\ -4 & -8 \end{vmatrix}$

12. $\begin{vmatrix} 3 & \frac{1}{2} \\ \frac{2}{3} & \frac{5}{6} \end{vmatrix}$

13. $\begin{vmatrix} 7 & -14 \\ 2 & -4 \end{vmatrix}$

14. $\begin{vmatrix} 3 & -2 \\ -4 & -6 \end{vmatrix}$

15. $\begin{vmatrix} 2i & 3i \\ 4i & -i \end{vmatrix}$

10.6 ♦ Solution by Determinants

We shall now consider the general form of a system of two linear equations and develop procedures that will serve as a model for any system. Let us consider the system

(1) $\quad a_1 x + b_1 y = c_1$
(2) $\quad a_2 x + b_2 y = c_2$

We shall solve this system using the analytic method discussed in Section 10.3.

Multiplying both members of Equation 1 by $-a_2$ and both members of Equation 2 by a_1 gives

(3) $\quad -a_1 a_2 x - a_2 b_1 y = -a_2 c_1$
(4) $\quad a_1 a_2 x + a_1 b_2 y = a_1 c_2$

Adding the members of Equations 3 and 4 gives

(5) $$a_1 b_2 y - a_2 b_1 y = a_1 c_2 - a_2 c_1$$
$$(a_1 b_2 - a_2 b_1) y = a_1 c_2 - a_2 c_1$$
$$y = \frac{a_1 c_2 - a_2 c_1}{a_1 b_2 - a_2 b_1} \qquad (a_1 b_2 - a_2 b_1 \neq 0)$$

Returning to the original system and multiplying both members of Equation 1 by b_2 and both members of Equation 2 by $-b_1$ gives

(6) $$a_1 b_2 x + b_1 b_2 y = b_2 c_1$$
(7) $$-a_2 b_1 x - b_1 b_2 y = -b_1 c_2$$

Adding the members of Equations 6 and 7 gives

(8) $$a_1 b_2 x - a_2 b_1 x = b_2 c_1 - b_1 c_2$$
$$(a_1 b_2 - a_2 b_1) x = b_2 c_1 - b_1 c_2$$
$$x = \frac{b_2 c_1 - b_1 c_2}{a_1 b_2 - a_2 b_1} \qquad (a_1 b_2 - a_2 b_1 \neq 0)$$

We now have the solution of the system:

$$\left(\frac{b_2 c_1 - b_1 c_2}{a_1 b_2 - a_2 b_1}, \frac{a_1 c_2 - a_2 c_1}{a_1 b_2 - a_2 b_1} \right)$$

The reader should check that this is indeed a solution of the system.

In observing this one ordered pair, which is the only solution of the given system, we see that the denominator in both of the components is the value of the determinant

$$\begin{vmatrix} a_1 & b_1 \\ a_2 & b_2 \end{vmatrix}$$

which we designate as D. The numerator of the expression for the x-component is the value of the determinant

$$\begin{vmatrix} c_1 & b_1 \\ c_2 & b_2 \end{vmatrix}$$

which we designate as D_x. The numerator of the expression for the y-component is the value of the determinant

$$\begin{vmatrix} a_1 & c_1 \\ a_2 & c_2 \end{vmatrix}$$

which we designate D_y.

10.6 SOLUTION BY DETERMINANTS

We can now write the solution of the system

(1) $$a_1x + b_1y = c_1$$
(2) $$a_2x + b_2y = c_2$$

as the ordered pair (x, y) where

$$x = \frac{D_x}{D} = \frac{\begin{vmatrix} c_1 & b_1 \\ c_2 & b_2 \end{vmatrix}}{\begin{vmatrix} a_1 & b_1 \\ a_2 & b_2 \end{vmatrix}} \quad \text{and} \quad y = \frac{D_y}{D} = \frac{\begin{vmatrix} a_1 & c_1 \\ a_2 & c_2 \end{vmatrix}}{\begin{vmatrix} a_1 & b_1 \\ a_2 & b_2 \end{vmatrix}}$$

Observe that the elements in the determinant D are the coefficients of the variables in equations 1 and 2:

$$D = \begin{vmatrix} a_1 & b_1 \\ a_2 & b_2 \end{vmatrix} \quad \begin{matrix} a_1x + b_1y = c_1 \\ a_2x + b_2y = c_2 \end{matrix}$$

The elements in determinant D_x result from the replacement in D of the coefficients of x in equations 1 and 2 by the constant terms of the two equations. The elements in determinant D_y result from the replacement in D of the coefficients of y in equations 1 and 2 by the constant terms of the two equations.

EXAMPLE. Find the solution set of

$$2x - 3y = -1$$
$$x + 4y = 5$$

SOLUTION. The coefficients of x are 2 and 1, respectively. The coefficients of y are -3 and 4, respectively. The constant terms are -1 and 5, respectively. Thus we have

$$D = \begin{vmatrix} 2 & -3 \\ 1 & 4 \end{vmatrix} = (2)(4) - (1)(-3) = 11$$

$$D_x = \begin{vmatrix} -1 & -3 \\ 5 & 4 \end{vmatrix} = (-1)(4) - (5)(-3) = 11$$

$$D_y = \begin{vmatrix} 2 & -1 \\ 1 & 5 \end{vmatrix} = (2)(5) - (1)(-1) = 11$$

Therefore

$$x = \frac{D_x}{D} = \frac{11}{11} = 1$$

and

$$y = \frac{D_y}{D} = \frac{11}{11} = 1$$

The solution set is $\{(1, 1)\}$.

If $D = 0$, this method is not valid because we cannot divide by zero. It can be shown that if $D = D_x = D_y = 0$, then the equations of the system are dependent. It can also be shown that if $D = 0$ and either D_x or D_y is not zero, the equations of the system are inconsistent.

EXERCISES 10.5

Find the solution set of each system by using determinants.

1. $x + 2y = 11$
   $3x - y = 5$
2. $3x + y = 7$
   $7x - y = 2$
3. $3x + 3y = 6$
   $4x + 6y = 7$
4. $x + y = 1$
   $6x - 3y = 0$
5. $5x - y = 10$
   $x + y = 8$
6. $x + 3y = 2$
   $2x - y = 10$
7. $x - y = 10$
   $3x - 2y = 1$
8. $4x - y = 0$
   $4x - 3y = -4$
9. $x - 3y = 2$
   $2x + y = 11$
10. $x - y = 2$
    $3x - 2y = 10$
11. $3x - y = -7$
    $x + 4y = 2$
12. $x - 4y = 1$
    $2x + y = -5$
13. $3x + 2y = 14$
    $2x - 3y = -21$
14. $4x - 3y = 12$
    $x + y = 3$
15. $2x - y = -6$
    $x + y = -4$
16. $x + y = 0$
    $-x + 2y = 3$
17. $3x + 2y = 0$
    $2x + y = 1$
18. $x + y = 2$
    $2x + 3y = 8$
19. $x - 2y = 1$
    $2x - y = 5$
20. $x + y = 4$
    $2x + y = 5$
21. $x - 2y = 8$
    $2x + y = -1$
22. $x - y = 0$
    $2x - y = -2$
23. $2x + y = -3$
    $3x - 2y = -22$
24. $-x + 2y = 7$
    $4x + 3y = -6$
25. $x + y = 0$
    $4x + 8y = -1$
26. $4x + y = 0$
    $6x - y = 5$
27. $x - 6y = -5$
    $4x + 3y = -12$
28. $x + 2y = 3$
    $4x + 8y = 12$

10.7 ANALYTIC SOLUTION OF A SYSTEM OF EQUATIONS IN THREE VARIABLES

29. $2x + y = 1$
    $3x - 2y = -9$
30. $x - 2y = 11$
    $2x - 3y = 18$
31. $6x + 2y = 11$
    $4x + 3y = 14$
32. $3x - 2y = 9$
    $3x - 5y = 18$
33. $2x - 3y = 20$
    $3x + 5y = 11$
34. $7x + 3y = 4$
    $14x + 9y = 7$
35. $4x - y = -1$
    $8x - 3y = -7$
36. $5x - 6y = 6$
    $10x - y = 10$
37. $4x + 2y = 7$
    $2x + y = 3$
38. $2x + 5y = 3$
    $3x - 2y = 14$
39. $\frac{2}{3}x + \frac{4}{3}y = 1$
    $\frac{2}{3}x - \frac{1}{4}y = \frac{7}{12}$
40. $x - y + 4 = 0$
    $\frac{1}{2}x + \frac{1}{6}y = -\frac{2}{5}$

10.7 ◆ Analytic Solution of a System of Equations in Three Variables

The analytic method used to solve a system of two linear equations in two variables may be used in solving a system of three linear equations in three variables:

$$a_1 x + b_1 y + c_1 z = d_1$$
$$a_2 x + b_2 y + c_2 z = d_2$$
$$a_3 x + b_3 y + c_3 z = d_3$$

A solution of the system above is an **ordered triple** of numbers (x, y, z), which satisfies the three equations in the system. We reduce this system to one of two linear equations in two variables.

EXAMPLE. Find the solution set of

(1) $\quad x + 2y + 3z = 14$
(2) $\quad 3x - 2y - z = -4$
(3) $\quad 2x - y + 3z = 9$

SOLUTION. We add the members of Equations 1 and 2 to get

(4) $\quad\quad\quad\quad 4x + 2z = 10$

We multiply the members of Equation 3 by 2 to get

(5) $\quad\quad\quad\quad 4x - 2y + 6z = 18$

We add the members of Equations 1 and 5 to get

(6) $\quad\quad\quad\quad 5x + 9z = 32$

The system of Equations 4 and 6, respectively,

$$4x + 2z = 10$$
$$5x + 9z = 32$$

is a system of two equations in two variables. Multiplying the members of Equation 4 by 5 and the members of Equation 6 by 4 gives

(7) $\qquad 20x + 10z = 50$
(8) $\qquad 20x + 36z = 128$

Subtracting the members of Equation 8 from the members of Equation 7, we have

(9) $\qquad -26z = -78$

which is equivalent to

$$z = 3$$

Replacing z by 3 in Equation 4 gives

(10) $\qquad 4x + 2(3) = 10$
$\qquad\qquad 4x = 4$
$\qquad\qquad x = 1$

Replacing x by 1 and z by 3 in Equation 1 gives

(11) $\qquad 1 + 2y + 3(3) = 14$
$\qquad\qquad 2y = 4$
$\qquad\qquad y = 2$

The solution set is $\{(1, 2, 3)\}$.

CHECK

$1 + 2(2) + 3(3) = 14;\qquad 3(1) - 2(2) - 3 = -4;\qquad 2(1) - 2 + 3(3) = 9$
$1 + 4 + 9 = 14;\qquad\qquad 3 - 4 - 3 = -4;\qquad\qquad 2 - 2 + 9 = 9$

EXERCISES 10.6

Find the solution sets and check.

1. $2x - y + z = 3$
   $3x + y - z = -8$
   $x - 2y + 3z = 11$
2. $x - 2y + z = 2$
   $2x - y + 3z = 2$
   $3x + y - z = 14$
3. $4x - 2y + z = 4$
   $2x + 4y - 2z = -3$
   $2x + 3z = 0$
4. $4x + 3y - z = 3$
   $3x - 2y + 2z = 5$
   $4y - 2z = -2$
5. $2x - 3y + 4z = 1$
   $3x + 6y - 2z = 3$
   $4x - 6y - 8z = -2$
6. $2x + z = 7$
   $y - z = -2$
   $x + y = 2$

10.8 ♦ Third Order Determinants

A determinant of the form

$$\begin{vmatrix} a_1 & b_1 & c_1 \\ a_2 & b_2 & c_2 \\ a_3 & b_3 & c_3 \end{vmatrix}$$

where $a_1, a_2, a_3, b_1, b_2, b_3, c_1, c_2,$ and c_3 are real numbers is called a 3×3 (read: "3 by 3") or **third-order** determinant.

A third-order determinant names a number. The value of this number is defined to be

$$\begin{vmatrix} a_1 & b_1 & c_1 \\ a_2 & b_2 & c_2 \\ a_3 & b_3 & c_3 \end{vmatrix} = a_1b_2c_3 - a_1b_3c_2 + a_2b_3c_1 - a_2b_1c_3 + a_3b_1c_2 - a_3b_2c_1$$

By a suitable factoring of pairs of terms in this definition, we obtain

$$\begin{vmatrix} a_1 & b_1 & c_1 \\ a_2 & b_2 & c_2 \\ a_3 & b_3 & c_3 \end{vmatrix} = a_1(b_2c_3 - b_3c_2) + a_2(b_3c_1 - b_1c_3) + a_3(b_1c_2 - b_2c_1)$$

$$= a_1(b_2c_3 - b_3c_2) - a_2(b_1c_3 - b_3c_1) + a_3(b_1c_2 - b_2c_1)$$

$$= a_1 \begin{vmatrix} b_2 & c_2 \\ b_3 & c_3 \end{vmatrix} - a_2 \begin{vmatrix} b_1 & c_1 \\ b_3 & c_3 \end{vmatrix} + a_3 \begin{vmatrix} b_1 & c_1 \\ b_2 & c_2 \end{vmatrix}$$

The 2×2 determinants in this sum are called the **minors** of their coefficients. The **minor** of any element of a determinant is defined to be the determinant that remains after the deletion of the row and column containing that element. In the 3×3 determinant

$$\begin{vmatrix} a_1 & b_1 & c_1 \\ a_2 & b_2 & c_2 \\ a_3 & b_3 & c_3 \end{vmatrix}$$

the minor of the element a_1 is

$$\begin{vmatrix} b_2 & c_2 \\ b_3 & c_3 \end{vmatrix}$$

the minor of the element b_2 is

$$\begin{vmatrix} a_1 & c_1 \\ a_3 & c_3 \end{vmatrix}$$

the minor of the element c_2 is

$$\begin{vmatrix} a_1 & b_1 \\ a_3 & b_3 \end{vmatrix}$$

and so forth.

By another factoring of pairs of terms in the definition of the value of the number named by the third-order determinant above, we see that

$$a_1(b_2c_3 - b_3c_2) - b_1(a_2c_3 - a_3c_2) + c_1(a_2b_3 - a_3b_2)$$

$$= a_1 \begin{vmatrix} b_2 & c_2 \\ b_3 & c_3 \end{vmatrix} - b_1 \begin{vmatrix} a_2 & c_2 \\ a_3 & c_3 \end{vmatrix} + c_1 \begin{vmatrix} a_2 & b_2 \\ a_3 & b_3 \end{vmatrix}$$

The sums above are examples of the **expansion** of a 3×3 determinant by the minors of the elements of the first column and of the first row, respectively. With the proper use of signs, it is possible to expand a determinant by the minors of the elements of *any* row or *any* column and obtain an expression for the value of the determinant. A helpful device for finding the signs of the terms in an expansion of a third-order determinant by minors is the array of alternating signs,

$$\begin{matrix} + & - & + \\ - & + & - \\ + & - & + \end{matrix}$$

which is called the **sign array** for the determinant. To obtain an expansion of a given third-order determinant about a given row or column, the appropriate sign from the sign array is prefixed to each term in the expansion. For example, if we expand the determinant

$$\begin{vmatrix} 1 & 2 & 3 \\ 0 & -1 & 2 \\ -2 & -1 & 1 \end{vmatrix}$$

about its first column, that is, by the minors of the elements of the first column, we have

10.8 THIRD ORDER DETERMINANTS

$$\begin{vmatrix} 1 & 2 & 3 \\ 0 & -1 & 2 \\ -2 & -1 & 1 \end{vmatrix} = (1)\begin{vmatrix} -1 & 2 \\ -1 & 1 \end{vmatrix} - (0)\begin{vmatrix} 2 & 3 \\ -1 & 1 \end{vmatrix} + (-2)\begin{vmatrix} 2 & 3 \\ -1 & 2 \end{vmatrix}$$

$$= (1)(-1+2) - 0 + (-2)(4+3)$$
$$= 1 - 0 - 14$$
$$= -13$$

If we expand the same determinant about the second row, that is by the minors of the elements of the second row, we have

$$\begin{vmatrix} 1 & 2 & 3 \\ 0 & -1 & 2 \\ -2 & -1 & 1 \end{vmatrix} = -(0)\begin{vmatrix} 2 & 3 \\ -1 & 1 \end{vmatrix} + (-1)\begin{vmatrix} 1 & 3 \\ -2 & 1 \end{vmatrix} - (2)\begin{vmatrix} 1 & 2 \\ -2 & -1 \end{vmatrix}$$

$$= 0 + (-1)(1+6) - 2(-1+4)$$
$$= -7 - 6$$
$$= -13$$

Note that the same value is obtained in each expansion.

EXERCISES 10.7

Evaluate by expanding about any row or column.

1. $\begin{vmatrix} 3 & -1 & 1 \\ 2 & -6 & 0 \\ 1 & 4 & 2 \end{vmatrix}$

2. $\begin{vmatrix} 3 & 1 & 6 \\ 1 & -2 & 3 \\ 2 & 4 & 2 \end{vmatrix}$

3. $\begin{vmatrix} 3 & 1 & -1 \\ 2 & 3 & -2 \\ 5 & 4 & 0 \end{vmatrix}$

4. $\begin{vmatrix} 5 & -1 & -1 \\ 2 & 3 & -2 \\ 3 & -2 & 1 \end{vmatrix}$

5. $\begin{vmatrix} 2 & 0 & 0 \\ 0 & 4 & 5 \\ 0 & 5 & 7 \end{vmatrix}$

6. $\begin{vmatrix} 1 & 0 & 0 \\ 0 & 1 & 0 \\ 0 & 0 & 1 \end{vmatrix}$

7. $\begin{vmatrix} 3 & 0 & -1 \\ 2 & 4 & 3 \\ 1 & -2 & 5 \end{vmatrix}$

8. $\begin{vmatrix} -1 & 2 & 0 \\ 3 & 0 & 2 \\ 4 & 4 & 0 \end{vmatrix}$

9. $\begin{vmatrix} 4 & 1 & 2 \\ 3 & 2 & 6 \\ -1 & 2 & 5 \end{vmatrix}$

10. $\begin{vmatrix} 2 & 3 & 12 \\ 4 & 2 & -2 \\ -3 & 4 & -1 \end{vmatrix}$

11. $\begin{vmatrix} a & a & a \\ 1 & 2 & 1 \\ 3 & 1 & 2 \end{vmatrix}$

12. $\begin{vmatrix} 0 & a & 0 \\ a & 0 & a \\ 0 & a & 0 \end{vmatrix}$

13. $\begin{vmatrix} x & y & 0 \\ x & 0 & y \\ 0 & x & y \end{vmatrix}$

14. $\begin{vmatrix} x & 0 & x \\ 0 & y & 0 \\ 0 & 0 & z \end{vmatrix}$

15. $\begin{vmatrix} 1 & 0 & 0 \\ 0 & x & 0 \\ 0 & 0 & x \end{vmatrix}$

16. $\begin{vmatrix} 0 & a & b \\ a & a & 0 \\ 0 & 0 & b \end{vmatrix}$

17. $\begin{vmatrix} x & y & z \\ 1 & 1 & 0 \\ 0 & 0 & 1 \end{vmatrix}$

18. $\begin{vmatrix} x & 0 & 0 \\ 1 & 2 & 3 \\ x & 0 & 0 \end{vmatrix}$

19. $\begin{vmatrix} x & y & 1 \\ x & y & 1 \\ 1 & 1 & 1 \end{vmatrix}$

20. $\begin{vmatrix} x^2 & y^2 & z^2 \\ x & y & z \\ x & y & z \end{vmatrix}$

10.9 ◆ Equations in Three Variables—Solutions by Determinants

Consider the system of three equations in three variables:

(1) $a_1x + b_1y + c_1z = d_1$
(2) $a_2x + b_2y + c_2z = d_2$
(3) $a_3x + b_3y + c_3z = d_3$

10.9 EQUATIONS IN THREE VARIABLES—SOLUTIONS BY DETERMINANTS

We can find the solution set of this system by methods demonstrated in Section 10.4. The ordered triple (x, y, z), which satisfies the system, is given by

$$x = \frac{D_x}{D}, \quad y = \frac{D_y}{D}, \quad z = \frac{D_z}{D}$$

where

$$D = \begin{vmatrix} a_1 & b_1 & c_1 \\ a_2 & b_2 & c_2 \\ a_3 & b_3 & c_3 \end{vmatrix} \neq 0$$

and

$$D_x = \begin{vmatrix} d_1 & b_1 & c_1 \\ d_2 & b_2 & c_2 \\ d_3 & b_3 & c_3 \end{vmatrix}, \quad D_y = \begin{vmatrix} a_1 & d_1 & c_1 \\ a_2 & d_2 & c_2 \\ a_3 & d_3 & c_3 \end{vmatrix}, \quad \text{and} \quad D_z = \begin{vmatrix} a_1 & b_1 & d_1 \\ a_2 & b_2 & d_2 \\ a_3 & b_3 & d_3 \end{vmatrix}$$

Observe that the elements of the determinant D are the coefficients of the variables in Equations 1, 2, and 3 and that D_x, D_y, and D_z are formed from D by replacing the elements that are the coefficients of x, y, and z, respectively, by the constant terms: d_1, d_2, and d_3.

EXAMPLE. Find the solution set of

$$\begin{aligned} x - 2y - z &= -2 \\ 2x + y + z &= 16 \\ x + 3y - 2z &= -2 \end{aligned}$$

SOLUTION. We have, by expanding the determinant about the first column:

$$\begin{aligned} D = \begin{vmatrix} 1 & -2 & -1 \\ 2 & 1 & 1 \\ 1 & 3 & -2 \end{vmatrix} \\ = (1)\begin{vmatrix} 1 & 1 \\ 3 & -2 \end{vmatrix} - (2)\begin{vmatrix} -2 & -1 \\ 3 & -2 \end{vmatrix} + (1)\begin{vmatrix} -2 & -1 \\ 1 & 1 \end{vmatrix} \\ = (1)(-5) - 2(7) + (1)(-1) \\ = -5 - 14 - 1 \\ = -20 \end{aligned}$$

Expanding D_x about the third column we have,

$$D_x = \begin{vmatrix} -2 & -2 & -1 \\ 16 & 1 & 1 \\ -2 & 3 & -2 \end{vmatrix} = (-1)\begin{vmatrix} 16 & 1 \\ -2 & 3 \end{vmatrix} - (1)\begin{vmatrix} -2 & -2 \\ -2 & 3 \end{vmatrix}$$

$$+ (-2)\begin{vmatrix} -2 & -2 \\ 16 & 1 \end{vmatrix}$$

$$= (-1)(50) - (1)(-10) + (-2)(30)$$
$$= -50 + 10 - 60$$
$$= -100$$

Expanding D_y and D_z about the first column we have,

$$D_y = \begin{vmatrix} 1 & -2 & -1 \\ 2 & 16 & 1 \\ 1 & -2 & -2 \end{vmatrix} = (1)\begin{vmatrix} 16 & 1 \\ -2 & -2 \end{vmatrix} - (2)\begin{vmatrix} -2 & -1 \\ -2 & -2 \end{vmatrix}$$

$$+ (1)\begin{vmatrix} -2 & -1 \\ 16 & 1 \end{vmatrix}$$

$$= (1)(-30) - (2)(2) + (1)(14)$$
$$= -30 - 4 + 14$$
$$= -20$$

$$D_z = \begin{vmatrix} 1 & -2 & -2 \\ 2 & 1 & 16 \\ 1 & 3 & -2 \end{vmatrix} = (1)\begin{vmatrix} 1 & 16 \\ 3 & -2 \end{vmatrix} - (2)\begin{vmatrix} -2 & -2 \\ 3 & -2 \end{vmatrix}$$

$$+ (1)\begin{vmatrix} -2 & -2 \\ 1 & 16 \end{vmatrix}$$

$$= (1)(-50) - (2)(10) + (1)(-30)$$
$$= -50 - 20 - 30$$
$$= -100$$

Then

$$x = \frac{D_x}{D} = \frac{-100}{-20} = 5$$

$$y = \frac{D_y}{D} = \frac{-20}{-20} = 1$$

$$z = \frac{D_z}{D} = \frac{-100}{-20} = 5$$

The solution set is $\{(5, 1, 5)\}$.

EXERCISES 10.8

Find the solution sets by using determinants.

1. $x + 2y + z = 6$
   $2x - y + z = 4$
   $x + y - z = 0$
2. $x + y + z = 7$
   $2x - 3y + z = -3$
   $3x + 2y - 5z = 2$
3. $3x - 2y + z = 18$
   $x - y - z = 2$
   $2x - 3y - 4z = -8$
4. $x + y = 2$
   $2x - z = 1$
   $3y - 3z = -1$
5. $x - 2y + z = -1$
   $y - z = 1$
   $3x + y - 2z = 6$
6. $x + y - 2z = 3$
   $3x - y + z = 5$
   $3x + 3y - 2z = 9$
7. $3x - 2y + 5z = 6$
   $5x - 4y + z = -5$
   $4x - 4y + 3z = 0$
8. $2x + 2y + z = 1$
   $3x + 2y - z = -4$
   $x + 4y + 6z = 13$
9. $2x + y = 18$
   $3x - 2y - 5z = 38$
   $y + z = -1$
10. $2x - 3y + 4z = 1$
    $x + 2y - \frac{2}{3}z = 1$
    $3x - 2y + 4z = 2$

10.10 ◆ The Conics

The graph of a second degree equation in the two variables x and y is a **conic**. A conic is the intersection of a plane with a right circular cone. We studied one of the conics—**the parabola**—in Chapter 8. Figure 10.8 shows how the various conics are formed. We shall discuss some of the conics.

Let us consider the equation

$$x^2 + y^2 = 25$$

Solving this equation explicitly for y, we have

$$y = \pm\sqrt{25 - x^2}$$

Assigning values to x, we find the following ordered pairs in the solution set:

$$(-5, 0), (-4, 3), (-3, 4), (0, 5), (3, 4), (4, 3), (5, 0),$$
$$(-4, -3), (-3, -4), (0, -5), (3, -4), (4, -3)$$

Plotting these points and drawing a smooth curve connecting them, we have the graph—a **circle** with radius 5 and center at the origin (Figure 10.9). In general, the graph of any equation of the form

$$x^2 + y^2 = r^2 \qquad (r > 0)$$

is a circle with radius r and center at the origin.

SYSTEMS OF EQUATIONS

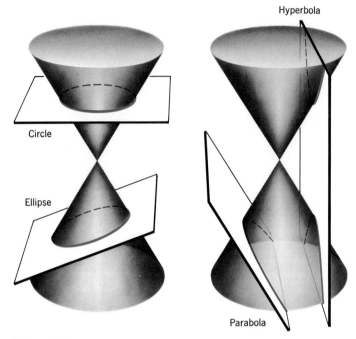

Figure 10.8

Note that in the previous example, values of x for which $|x| > 5$ need not be assigned because y is complex for these values.

Now let us consider the equation

$$9x^2 + 4y^2 = 36$$

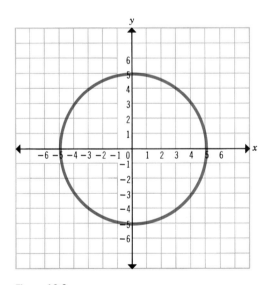

Figure 10.9

10.10 THE CONICS

Solving this equation explicitly for y, we obtain

$$y = \pm \sqrt{\frac{36 - 9x^2}{4}}$$
$$= \pm \tfrac{3}{2}\sqrt{4 - x^2}$$

Assigning numbers between -2 and 2 as values for x (note that when $x > 2$ or $x < -2$, y is complex), we obtain, for example,

$(0, 3)$, $\left(1, \frac{3\sqrt{3}}{2}\right)$, $(2, 0)$, $\left(1, -\frac{3\sqrt{3}}{2}\right)$, $(-2, 0)$,

$\left(-1, \frac{3\sqrt{3}}{2}\right)$, $\left(-1, -\frac{3\sqrt{3}}{2}\right)$, $(0, -3)$

Plotting these points and connecting them with a smooth curve, we have the graph in Figure 10.10. This curve is called an **ellipse.**

The graph of any equation of the form

$$Ax^2 + By^2 = C$$

where A, B, and C are positive real numbers and $A \neq B$, is an ellipse whose center is at the origin, with x-intercepts $\sqrt{\frac{C}{A}}$ *and* $-\sqrt{\frac{C}{A}}$, *and with y-intercepts* $\sqrt{\frac{C}{B}}$ *and* $-\sqrt{\frac{C}{B}}$.

Now let us consider the equation

$$9x^2 - 25y^2 = 225$$

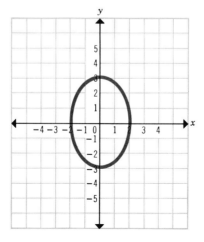

Figure 10.10

Solving this equation explicitly for y, we have

$$y = \pm \sqrt{\frac{9x^2 - 225}{25}}$$

$$= \pm \tfrac{3}{5}\sqrt{x^2 - 25}$$

We assign to x values that are greater than or equal to 5 or less than or equal to -5 (why?) and obtain, for example,

$(-7, \tfrac{6}{5}\sqrt{6}), (-7, -\tfrac{6}{5}\sqrt{6}), (-6, \tfrac{3}{5}\sqrt{11}), (-6, -\tfrac{3}{5}\sqrt{11}), (-5, 0),$
$(5, 0), (6, \tfrac{3}{5}\sqrt{11}), (6, -\tfrac{3}{5}\sqrt{11}), (7, \tfrac{6}{5}\sqrt{6}), (7, -\tfrac{6}{5}\sqrt{11}).$

Locating these points on the plane and connecting them by a smooth curve, we have the graph in Figure 10.11. This curve is called a **hyperbola**. *Any equation of the form*

$$Ax^2 - By^2 = C$$

where A, B, and C are positive real numbers, has a graph which is a hyperbola with center at the origin and x-intercepts $\sqrt{\dfrac{C}{A}}$ *and* $-\sqrt{\dfrac{C}{A}}$.

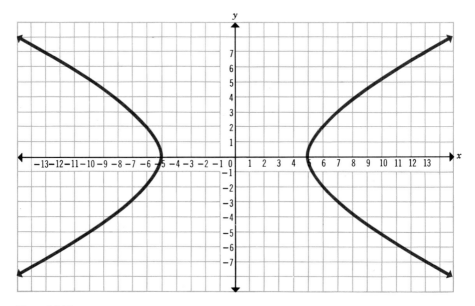

Figure 10.11

10.11 SYSTEMS OF ONE LINEAR AND ONE SECOND DEGREE EQUATION

EXERCISES 10.9

Name and sketch the graph of each of the following equations.

1. $x^2 + y^2 = 4$
2. $4x^2 + y^2 = 4$
3. $x^2 + y^2 = 1$
4. $x^2 - y^2 = 1$
5. $x^2 - y^2 = 9$
6. $x^2 + y^2 = 49$
7. $3x^2 + 3y^2 = 27$
8. $4x^2 + 9y^2 = 36$
9. $x^2 + 9y^2 = 9$
10. $4x^2 - 25y^2 = 100$
11. $5x^2 + 5y^2 = 45$
12. $25x^2 + 9y^2 = 225$
13. $4x^2 + 25y^2 = 100$
14. $y = 3x^2 - 4x + 9$
15. $y^2 = 144 - x^2$
16. $x^2 - y^2 = 36$
17. $x^2 - y^2 = 144$
18. $49x^2 + 4y^2 = 196$
19. $25x^2 - 4y^2 = 100$
20. $2x^2 + 2y^2 = 15$

10.11 ◆ Systems of One Linear and One Second Degree Equation

Let us consider the system

(1) $\qquad x^2 + y^2 = 17$
(2) $\qquad x + y = 3$

The graph of Equation 1 is a circle. The graph of Equation 2 is a line. We would like to find the solution set of this system; that is, we would like to find the set of all ordered pairs that satisfy both equations in the system.

Let us solve Equation 2 explicitly for y. We have

$$y = 3 - x$$

Substituting this expression for y in Equation 1 we get

$$x^2 + (3 - x)^2 = 17$$

Simplifying, we have the following equivalent statements:

$$x^2 + 9 - 6x + x^2 = 17$$
$$2x^2 - 6x - 8 = 0$$
$$x^2 - 3x - 4 = 0$$
$$(x - 4)(x + 1) = 0$$
$$x - 4 = 0 \quad \text{or} \quad x + 1 = 0$$
$$x = 4 \quad \text{or} \quad x = -1$$

When we replace x by 4 in the equation $y = 3 - x$, we have $y = -1$. When we replace x by -1 in the same equation we have $y = 4$.

The solution set is $\{(4, -1), (-1, 4)\}$.

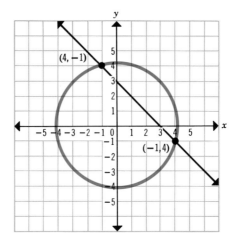

Figure 10.12

Checking, we have

$$(4)^2 + (-1)^2 = 17 \qquad (-1)^2 + (4)^2 = 17$$
$$16 + 1 = 17 \qquad 1 + 16 = 17$$
$$4 = 3 - (-1) \qquad -1 = 3 - 4$$
$$4 = 3 + 1 \qquad -1 = -1$$

The graph of this system is shown in Figure 10.12.

The reason behind the substitution used in solving the above system is this: If an ordered pair exists that satisfies both equations in a system, for the components of that ordered pair the variables of any one equation of the system may be considered, as entities, respectively equal and interchangeable with the variables of the other equation. Note that we said IF an ordered pair exists that satisfies both equations of the system. It may be that the ordered pair is a solution of one of the equations of the system and not of the other. Whether the ordered pair found is a solution of the system must be decided by replacing the variables of the equations in the system with these values to see if true statements result in each case.

10.12 ◆ Systems of Two Second-Degree Equations

Finding the solution set of a system of two second-degree equations may become very involved. We shall illustrate by examples some methods for solving such systems.

10.12 SYSTEMS OF TWO SECOND-DEGREE EQUATIONS

EXAMPLE 1. Find the solution set of the system

(1) $$x^2 - y^2 = 7$$
(2) $$2x^2 + 3y^2 = 24$$

SOLUTION. Multiplying both members of Equation 1 by 3, we have

(3) $$3x^2 - 3y^2 = 21$$

Adding the members of Equations 2 and 3 produces

(4) $$5x^2 = 45$$

which is equivalent to

$$x^2 = 9$$
$$x = 3 \quad \text{or} \quad x = -3$$

Replacing x by 3 in Equation 1 gives

$$(3)^2 - y^2 = 7$$

which is equivalent to

$$9 - y^2 = 7$$
$$y^2 = 2$$
$$y = \sqrt{2} \quad \text{or} \quad y = -\sqrt{2}$$

Thus we find that the two ordered pairs, $(3, \sqrt{2})$ and $(3, -\sqrt{2})$, are solutions of this system.

If we replace x by -3 in Equation 1, we have the same sequence of open sentences as above and we find that the ordered pairs $(-3, \sqrt{2})$ and $(-3, -\sqrt{2})$ are also solutions of the system.

The solution set is $\{(-3, \sqrt{2}), (-3, -\sqrt{2}), (3, \sqrt{2}), (3, -\sqrt{2})\}$.

EXAMPLE 2. Find the solution set of the system

(1) $$x^2 - xy + y^2 = 7$$
(2) $$x^2 + y^2 = 5$$

SOLUTION. We subtract the members of Equation 2 from the members of Equation 1 to obtain

(3) $$-xy = 2$$

which is equivalent to

(4) $$y = -\frac{2}{x} \quad x \neq 0$$

Replacing y by $-\dfrac{2}{x}$ in Equation 2, we have

$$x^2 + \left(-\dfrac{2}{x}\right)^2 = 5 \qquad x \neq 0$$

$$x^2 + \dfrac{4}{x^2} = 5 \qquad x \neq 0$$

$$x^4 + 4 = 5x^2 \qquad x \neq 0$$

$$x^4 - 5x^2 + 4 = 0 \qquad x \neq 0$$

Treating this equation as one in quadratic form, we have

$$(x^2)^2 - 5(x^2) + 4 = 0 \qquad x \neq 0$$
$$(x^2 - 4)(x^2 - 1) = 0 \qquad x \neq 0$$
$$x^2 = 4 \quad \text{or} \quad x^2 = 1$$
$$x = 2, \; x = -2, \; x = 1, \text{ or } x = -1$$

When we replace x by 2 in Equation 4, we get $y = -1$. When we replace x by -2 in Equation 4, we get $y = 1$. When we replace x by 1 in Equation 4, we get $y = -2$. When we replace x by -1 in Equation 4, we get $y = 2$.

The solution set is $\{(2, -1), (-2, 1), (1, -2), (-1, 2)\}$.

EXERCISES 10.10

Find the solution sets of the following systems.

1. $x^2 - y^2 = 16$
   $x + 3y = 4$
2. $x^2 + y^2 = 16$
   $y = 5$
3. $y^2 = 4x$
   $y = -x$
4. $x^2 - y^2 = 16$
   $x - 3y = 0$
5. $2x^2 + y^2 = 24$
   $x - y = 0$
6. $x^2 + 2y^2 = 16$
   $\sqrt{2}x + 2y - 4\sqrt{2} = 0$
7. $x^2 + y^2 = 4$
   $x = 2 - y$
8. $x^2 - y^2 = 16$
   $3x - y = 12$
9. $y = 3x^2$
   $x = y$
10. $x^2 + y^2 = 25$
    $x^2 - y^2 = 7$
11. $x^2 + y^2 = 5$
    $4x^2 + 9y^2 = 25$
12. $x^2 + y^2 = 25$
    $3x^2 - y^2 = 39$
13. $x^2 + y^2 = 100$
    $x^2 + 2y^2 = 164$
14. $x^2 - 3y = 0$
    $y^2 - x^2 = 108$
15. $x^2 - 10y^2 = 16$
    $x^2 - 4y^2 = 20$
16. $9x^2 + y^2 = 18$
    $x^2 + y^2 = 10$

17. $x^2 + y^2 = 25$
    $x^2 + 4y^2 = 52$
18. $x^2 - y^2 = 3$
    $3x^2 + 4y^2 = 16$
19. $x^2 + 4y^2 = 67$
    $4x^2 - 9y^2 = -132$
20. $x^2 + y^2 = 9$
    $5x^2 + 5y^2 = 25$

CHAPTER 10 REVIEW

Find the solution sets by graphing (Exercises 1–2).

1. $2x - y = 4$
   $x + y = 5$
2. $3x + 4y = -2$
   $-2x + y = 5$

Find the solution sets analytically (Exercises 3–4).

3. $3x + 2y = 6$
   $2x + y = 5$
4. $x + y = 1$
   $2x + 3y = 0$

Find the number named by the following determinants (Exercises 5–6).

5. $\begin{vmatrix} -1 & 3 \\ -2 & 4 \end{vmatrix}$
6. $\begin{vmatrix} -4 & 3 \\ -2 & -1 \end{vmatrix}$

Find the solution sets by using determinants (Exercises 7–8).

7. $3x + 2y = 5$
   $2x - 5y = -22$
8. $3x + 2y = 1$
   $2x - y = 3$

Find the solution sets analytically (Exercises 9–10).

9. $x + y + z = 0$
   $2x - y + 3z = 2$
   $3x - 2y + z = 5$
10. $x - y - 2z = 7$
    $2x + y - z = 2$
    $3x + 2y + 4z = -4$

Find the number named by the following determinants (Exercises 11–12).

11. $\begin{vmatrix} 1 & -1 & 2 \\ 0 & 3 & -4 \\ 2 & 1 & 3 \end{vmatrix}$
12. $\begin{vmatrix} -1 & 3 & 4 \\ 2 & 0 & -2 \\ 1 & 7 & 1 \end{vmatrix}$

Use determinants to find the solution sets (Exercises 13–14).

13. $2x + y + z = 1$
    $x + 2y - 3z = 0$
    $3x + y + 2z = 1$
14. $5x - 2y + z = 5$
    $4x + y + 3z = 1$
    $7x - 3y - 2z = 4$

Sketch the graphs of the following equations (Exercises 15–17).

15. $x^2 + y^2 = 9$ 16. $4x^2 + 9y^2 = 36$ 17. $25x^2 - 9y^2 = 225$

Find the solution sets (Exercises 18–20).

18. $x^2 + y^2 = 25$
    $x - y = 1$

19. $9x^2 + 4y^2 = 36$
    $3x + 2y = 6$

20. $x^2 + y^2 = 25$
    $xy = 12$

CHAPTER 11

Sequences and Series

11.1 ◆ Sequences

A **sequence** is an ordered set of numbers in one-to-one correspondence with the natural numbers:

$$a_1, a_2, a_3, \ldots, a_n, \ldots$$

Each number in the sequence is called a **term** of the sequence. The first term is denoted by a_1, the second by a_2, and so on. The nth term, which is called the **general term** of the sequence, is denoted by a_n. Some examples of sequences are:

$$1, 3, 5, 7, 9, \ldots, 2n - 1, \ldots$$
$$2, 4, 6, 8, 10, \ldots, 2n, \ldots$$
$$1, 8, 27, 64, \ldots, n^3, \ldots$$

A sequence with a finite number of terms is called a **finite sequence** and one with an infinite number of terms is called an **infinite sequence.**

The set of ordered pairs $\{(1, a_1), (2, a_2), \ldots (n, a_n), \ldots\}$ where $a_1, a_2, \ldots, a_n, \ldots$ are terms of a sequence is a special kind of function called a **sequence function.** The domain of a sequence function is the set of natural numbers; its range is the set of elements of the sequence.

EXAMPLE 1. Find the first four terms of the sequence whose general term is n^2.

SOLUTION. We find the terms in a sequence by replacing n in the general term by the natural numbers in turn. Thus, for the given sequence, we have

$$a_1 = 1^2 = 1$$
$$a_2 = 2^2 = 4$$
$$a_3 = 3^2 = 9$$
$$a_4 = 4^2 = 16$$

When we are given the first few terms of a sequence, it may be possible to determine an expression for the general term of the sequence to which these terms belong. Sometimes, determining the general term of a sequence is quite easy; at other times, it is difficult. It should be observed that the general term of a sequence is not uniquely determined. For example, if the first three terms of a sequence are

$$2, 4, 6$$

the general term may be

$$a_n = 2n$$

or

$$a_n = 2n + (n-1)(n-2)(n-3)$$

Using both formulas given above for a_n, we generate 2, 4, 6 for the first three terms of the sequence. Notice that in the case $a_n = 2n$, $a_4 = 8$, while in the case $a_n = 2n + (n-1)(n-2)(n-3)$, $a_4 = 14$.

EXAMPLE 2. Find the first five terms of the sequence whose nth term is $(n+1)(2n+1)$.

SOLUTION. Replacing n in the formula by 1, 2, 3, 4, and 5 in turn, we have

$$a_1 = (1+1)(2 \cdot 1 + 1) = 6$$
$$a_2 = (2+1)(2 \cdot 2 + 1) = 15$$
$$a_3 = (3+1)(2 \cdot 3 + 1) = 28$$
$$a_4 = (4+1)(2 \cdot 4 + 1) = 45$$
$$a_5 = (5+1)(2 \cdot 5 + 1) = 66$$

EXERCISES 11.1

Find the first four terms of a sequence with the given general term (Exercises 1–20).

1. $a_n = \dfrac{n(n+1)}{2}$ 2. $a_n = \dfrac{3n}{2}(n+1)$

3. $a_n = n^2$
4. $a_n = \dfrac{n}{2n+1}$
5. $a_n = n + 3$
6. $a_n = 3n(n+1)$
7. $a_n = n - 4$
8. $a_n = 1 + (-1)^n$
9. $a_n = \dfrac{n}{4n+1}$
10. $a_n = 2^{2n}$
11. $a_n = \dfrac{n}{n+1}$
12. $a_n = 4^n - 1$
13. $a_n = (n+1)(n-1)$
14. $a_n = n(3n^2 + 6n + 1)$
15. $a_n = 2n^2 - 1$
16. $a_n = n^3 - 1$
17. $a_n = (n-1)^{2n}$
18. $a_n = n^{2n}$
19. $a_n = n^3 - n^2$
20. $a_n = 3n(n^2 + 1)$

Find a general term for the following sequences whose first four terms are given (Exercises 21–26).

21. $1, \frac{1}{2}, \frac{1}{3}, \frac{1}{4}$
22. 3, 6, 9, 12
23. $\frac{1}{2}, 0, -\frac{1}{2}, -1$
24. $1, \frac{1}{2}, \frac{1}{4}, \frac{1}{8}$
25. $1, -1, 1, -1$
26. a, ar, ar^2, ar^3

11.2 ◆ Series

Whenever we have a finite sequence

$$a_1, a_2, \ldots, a_n$$

the sum of all the terms of this sequence

$$a_1 + a_2 + \cdots + a_n$$

is called a **finite series.**

Thus, with the finite sequence,

$$2, 5, 8, \ldots, [2 + 3(n-1)]$$

is associated the finite series

$$2 + 5 + 8 + \cdots + [2 + 3(n-1)]$$

Similarly, with the finite sequence,

$$a, a^2, a^3, \ldots, a^n$$

is associated the finite series

$$a + a^2 + a^3 + \cdots + a^n$$

Since the terms in the finite series are the same as the terms of the sequence, we refer to the first term, second term, . . . , or the general term of the series in the same manner as we do for a sequence.

A highly useful symbol for denoting a finite series is the uppercase Greek letter sigma, Σ, used in **summation notation.** We denote the finite series

$$a_1 + a_2 + a_3 + \cdots + a_n$$

by

$$\sum_{i=1}^{n} a_i$$

Let us consider the finite sequence

$$3, 6, 9, 12, \ldots, 3n$$

The finite series associated with this sequence is denoted by

$$\sum_{i=1}^{n} 3i = 3\cdot 1 + 3\cdot 2 + 3\cdot 3 + \cdots + 3n$$
$$= 3 + 6 + 9 + \cdots + 3n$$

We read the symbol $\sum_{i=1}^{n} 3i$, "The summation from 1 to n of $3i$." When we see this symbol, it is understood that we are to add all terms obtained from $3i$ by replacing i with the consecutive natural numbers from 1 to n, inclusive. The symbol Σ is called the **summation sign.** The expression $3i$, in this case, is called the **summand** and the letter i, the **index.**

EXAMPLE 1. Write $\sum_{i=1}^{4} i(i-1)$ in expanded notation.

SOLUTION

$$\sum_{i=1}^{4} i(i-1) = 1(1-1) + 2(2-1) + 3(3-1) + 4(4-1)$$
$$= 0 + 2 + 6 + 12$$

EXAMPLE 2. Write $1^4 + 2^4 + 3^4 + 4^4 + 5^4 + 6^4$ using summation notation.

SOLUTION

$$1^4 + 2^4 + 3^4 + 4^4 + 5^4 + 6^4 = \sum_{i=1}^{6} i^4$$

EXERCISES 11.2

Write in expanded form (Exercises 1–10).

1. $\sum_{i=1}^{4} (3i + 1)$ 2. $\sum_{i=2}^{8} (2i - 3)$

3. $\sum_{i=1}^{4}(3^i - 1)$

4. $\sum_{i=2}^{6}(-1)^i 2^i$

5. $\sum_{i=1}^{7} \frac{i}{i+1}$

6. $\sum_{i=1}^{7} \frac{(-1)^i}{i+2}$

7. $\sum_{i=1}^{n}(3i + 4)$

8. $\sum_{i=1}^{n}(2^i - 1)$

9. $\sum_{i=1}^{n} \frac{(-1)^{i-1}}{2i}$

10. $\sum_{i=1}^{n}(-1)^i(3^i + 1)$

Write using summation notation (Exercises 11–20).

11. $1 + 2 + 3 + 4 + 5$

12. $a + a^2 + a^3 + a^4$

13. $1 + 3 + 5 + 7 + 9$

14. $3 + 6 + 9 + \cdots + 3n$

15. $\frac{1}{3} + \frac{1}{9} + \frac{1}{27} + \cdots + \frac{1}{3^n}$

16. $x - x^2 + x^3 - \cdots + (-1)^{n-1}x^n$

17. $\frac{1}{2} + \frac{1}{3} + \frac{1}{4} + \cdots + \frac{1}{n}$

18. $5 + 8 + 11 + \cdots + [3 + (3n - 1)]$

19. $2 - 2^2 + 2^3 - 2^4 + \cdots + (-1)^{n-1}2^n$

20. $\frac{1}{2} + \frac{2}{3} + \frac{3}{4} + \cdots + \frac{n}{n+1}$

11.3 ◆ Arithmetic Progressions

A sequence of numbers

$$a_1, a_2, a_3, \ldots, a_n$$

is called an **arithmetic progression** (accented ar ith met′ ic) if and only if there is some number, d, for which

$$a_2 = a_1 + d$$
$$a_3 = a_1 + 2d$$
$$a_4 = a_1 + 3d$$
$$\vdots$$
$$a_{n-1} = a_1 + (n - 2)d$$
$$a_n = a_1 + (n - 1)d$$

The number d is called the **common difference** of the progression. Some examples of arithmetic progressions are:

(1)   6, 10, 14, 18, …
(2)   3, 2, 1, 0, −1, …
(3)   3, 1, −1, −3, −5, …

The common difference in (1) is 4, in (2) is −1, and in (3) is −2.

We can verify that a finite sequence is an arithmetic progression by simply subtracting each term from its successor and noting that the difference in each case is the same. For example,

$$3, 8, 13, 18$$

is an arithmetic progression because

$$8 - 3 = 5$$
$$13 - 8 = 5$$
$$18 - 13 = 5$$

If at least three terms of an arithmetic progression are known, we can determine the common difference and generate as many terms as we wish. We can also find an expression for the general term. Consider the arithmetic progression with first term a and common difference d. Then:

The first term is a.
The second term is $a + d$.
The third term is $(a + d) + d = a + 2d$.
The fourth term is $(a + 2d) + d = a + 3d$.
$$\vdots$$
The nth term is $(a + [n - 2]d) + d = a + (n - 1)d$.

From this we see that

$$a_n = a + (n - 1)d$$

EXAMPLE 1. Find the tenth term of the arithmetic progression

$$1, -2, -5, \ldots$$

SOLUTION. The first term is 1, the common difference is $-5-(-2)=-3$, and the number of terms is 10. Thus the tenth term is

$$\begin{aligned} a_{10} &= 1 + (10 - 1)(-3) \\ &= 1 + 9(-3) \\ &= 1 - 27 \\ &= -26 \end{aligned}$$

CHECK

$$1, -2, -5, -8, -11, -14, -17, -20, -23, -26$$

EXAMPLE 2. The fourteenth term of an arithmetic progression is -15. The common difference is -5. Find the first term and write the first four terms of the arithmetic progression.

SOLUTION. Since -15 is the fourteenth term and -5 is the common difference, we have

$$-15 = a_1 + (14 - 1)(-5)$$
$$-15 = a_1 + (13)(-5)$$
$$-15 = a_1 - 65$$
$$a_1 = 50$$

The first four terms are 50, 45, 40, 35.

11.4 ◆ Arithmetic Series

Let us consider the series of n terms associated with the arithmetic progression

$$a, a + d, a + 2d, \ldots, [a + (n - 1)d]$$

that is,

(1) $\qquad S_n = a + (a + d) + (a + 2d) + \cdots + [a + (n - 1)d]$

We can also write this series in the form

(2) $\quad S_n = [a + (n - 1)d] + [a + (n - 2)d] + [a + (n - 3)d] + \cdots + a$

Adding the members of (1) and (2), term by term, gives

$$2S_n = [2a + (n - 1)d] + [2a + (n - 1)d] + \cdots + [2a + (n - 1)d]$$

where each term in the right member is $2a + (n - 1)d$ and there are n terms. Then

$$2S_n = n[2a + (n - 1)d]$$
$$S_n = \frac{n}{2}[2a + (n - 1)d]$$

This series is called an **arithmetic series**.

We may write

$$S_n = \frac{n}{2}\{a + [a + (n - 1)d]\}$$

$$S_n = \frac{n}{2}(a + a_n)$$

since $a_n = a + (n - 1)d$.

EXAMPLE 1. Find the sum of the finite arithmetic series

$$\sum_{i=1}^{10}(2i + 1)$$

SOLUTION. We write the first two or three terms in expanded form:

$$3 + 5 + 7 + \cdots$$

We can see that the first term is 3 and the common difference is 2. There are 10 terms. Thus, using the equation

$$S_n = \frac{n}{2}[2a + (n-1)d]$$

we have

$$\begin{aligned}
S_{10} &= \tfrac{10}{2}[2(3) + (10-1)2] \\
&= 5[6 + (9)(2)] \\
&= 5(6 + 18) \\
&= 5(24) \\
&= 120
\end{aligned}$$

EXAMPLE 2. Find the sum of the first one hundred natural numbers.

SOLUTION. The first few terms in the expanded form are

$$1 + 2 + 3 + \cdots$$

We can see that the first term is 1 and the common difference is 1. There are 100 terms. Thus, using the equation $S_n = \frac{n}{2}[2a + (n-1)d]$,

we have

$$\begin{aligned}
S_{100} &= \tfrac{100}{2}[(2)(1) + (100-1)(1)] \\
&= 50[2 + 99] \\
&= (50)(101) \\
&= 5050
\end{aligned}$$

EXERCISES 11.3

Find the common difference and the nth term of each of the following arithmetic progressions (Exercises 1–10).

1. 2, 4, 6, ...
2. 13, 17, 21, ...
3. −8, −3, 2 ...
4. 1, 0, −1, ...
5. 13, 8, 3, ...
6. 1000, 1001, 1002, ...
7. $a, a+1, a+2, \ldots$
8. $2z - 3, 2z - 1, 2z + 1, \ldots$
9. $\tfrac{1}{2}, \tfrac{1}{4}, 0, \ldots$
10. $-4, -2\tfrac{1}{2}, -1, \ldots$

Find the indicated term in each of the following arithmetic progressions (Exercises 11–16).

11. $-8, -3, 2, \ldots$; 18th
12. $2, 4, 6, \ldots$; 36th
13. $3, 10, 17, \ldots$; 48th
14. $19, 17, 15, \ldots$; 32nd
15. $-1, -3, -5, \ldots$; 10th
16. $a + 2b, 3a + 3b, 5a + 4b, \ldots$; 12th

Find the sums of the finite series below (Exercises 17–22).

17. $\sum_{i=1}^{4} (2i + 1)$
18. $\sum_{i=1}^{8} (2i - 5)$
19. $\sum_{i=1}^{20} (3i + 1)$
20. $\sum_{i=1}^{100} (2i - 1)$
21. $\sum_{i=1}^{50} 2i$
22. $\sum_{i=1}^{100} 3i$

23. The tenth term of an arithmetic progression is 32, and the eighteenth term is 48. Find the first term and the common difference.
24. The first term of an arithmetic progression is -5 and the tenth term is 13. Find the common difference.
25. Find the first term of an arithmetic progression whose third term is 7 and whose eighth term is 17.
26. In the arithmetic progression: $10, 6, 2, \ldots$, which term is -46?
27. The fourteenth term of an arithmetic progression is 72.5, and the twentieth term is 93.5. Find the sum of the first four terms.
28. A man is offered a position at a salary which starts at $6000 per year and increases yearly by $400. How much would his total earnings amount to if he worked ten years under this salary schedule?
29. On a construction project, a contractor was penalized for taking more than the contracted time to finish the job. He forfeits $60 the first day, $75 the second day, $90 the third day, and so on. How many extra days did he need to complete the project if he paid a penalty of $1080?

11.5 ◆ Geometric Progressions

Any sequence in which each term after the first is obtained by multiplying the preceding term by a fixed factor, called the **common ratio,** is called a **geometric progression.** The following sequences are examples of geometric progressions with the common ratio of each given in parentheses:

$$2, 4, 8, 16, \ldots \quad (2)$$
$$1, -\tfrac{1}{2}, \tfrac{1}{4}, -\tfrac{1}{8}, \ldots \quad (-\tfrac{1}{2})$$
$$8, 2, \tfrac{1}{2}, \ldots \quad (\tfrac{1}{4})$$

If the first term in a geometric progression is designated by a and the common ratio by r:

SEQUENCES AND SERIES

The second term is ar
The third term is $(ar)r = ar^2$
The fourth term is $(ar^2)r = ar^3$
$$\vdots$$
The nth term is $(ar^{n-2})r = ar^{n-1}$

Thus we can see that in a geometric progression,

$$a_n = ar^{n-1}$$

The general form of a geometric progression is

$$a, ar, ar^2, ar^3, \ldots, ar^{n-1}, \ldots$$

EXAMPLE 1. Find the eighth term in the geometric progression

$$\tfrac{1}{8}, \tfrac{1}{4}, \tfrac{1}{2}, \ldots$$

SOLUTION. By observation, we see that the first term is $\tfrac{1}{8}$ and the common ratio is 2. Then, using the formula

$$a_n = ar^{n-1}$$

with $n = 8$, we have

$$\begin{aligned} a_8 &= (\tfrac{1}{8})(2^7) \\ &= \tfrac{128}{8} \\ &= 16 \end{aligned}$$

EXAMPLE 2. In the geometric progression

$$32, 16, 8, \ldots$$

which term is $\tfrac{1}{64}$?

SOLUTION. We see that the first term is 32 and the common ratio is $\tfrac{1}{2}$. Using the formula for the nth term of a geometric progression, we have

$$\frac{1}{64} = (32)\left(\frac{1}{2}\right)^{n-1}$$

$$\frac{1}{(64)(32)} = \left(\frac{1}{2}\right)^{n-1}$$

$$\frac{1}{2^6 \cdot 2^5} = \frac{1}{2^{n-1}}$$

$$\frac{1}{2^{11}} = \frac{1}{2^{n-1}}$$

$$n - 1 = 11$$

$$n = 12$$

11.6 ◆ Geometric Series

The sum of the terms of the geometric progression with n terms

$$a, ar, ar^2, ar^3, \ldots, ar^{n-1}$$

is the **geometric series** given by

(1) $\qquad S_n = a + ar + ar^2 + ar^3 + \cdots + ar^{n-1}$

We would like to find an explicit expression for this sum in terms of a, r, and n. Multiplying both members of equation 1 by r gives

(2) $\qquad r \cdot S_n = ar + ar^2 + ar^3 + \cdots + ar^n$

Subtracting the members, term by term, of equation 2 from those of equation 1, we have

$$S_n - r \cdot S_n = a - ar^n$$
$$(1 - r)S_n = a(1 - r^n)$$
$$S_n = a\frac{1 - r^n}{1 - r}$$

We now have a formula for the sum of a geometric series with n terms. For the particular case in which $r = 1$, the formula is not valid (why?). For this case in which $r = 1$, we have the geometric series

$$a + a + a + \cdots + a$$

with n terms; the sum is obviously na.

An alternate expression for S_n can be obtained by noting that

$$S_n = a\frac{1 - r^n}{1 - r}$$

may be written as

$$S_n = \frac{a - ar^n}{1 - r}$$
$$= \frac{a - r(ar^{n-1})}{1 - r}$$

and, since $a_n = ar^{n-1}$, we obtain

$$S_n = \frac{a - ra_n}{1 - r}, \qquad r \neq 1$$

EXAMPLE 1. Find $\sum_{i=1}^{5} 2^i$.

SOLUTION. We write the first two or three terms in expanded form

$$2^1 + 2^2 + 2^3 + \cdots = 2 + 4 + 8 + \cdots$$

We see that the first term is 2 and the common ratio is 2. For $n = 5$ we have

$$\begin{aligned} S_5 &= 2 \cdot \frac{1 - 2^5}{1 - 2} \\ &= 2 \cdot \frac{1 - 32}{-1} \\ &= 2\left(\frac{-31}{-1}\right) \\ &= 2(31) \\ &= 62 \end{aligned}$$

EXAMPLE 2. Find $\sum_{i=3}^{12} 2^{i-5}$.

SOLUTION. We write the first two or three terms in expanded form:

$$2^{3-5} + 2^{4-5} + 2^{5-5} + \cdots$$
$$2^{-2} + 2^{-1} + 2^0 + \cdots$$
$$\frac{1}{2^2} + \frac{1}{2} + 1 + \cdots$$
$$\frac{1}{4} + \frac{1}{2} + 1 + \cdots$$

The first term is $\frac{1}{4}$ and the common ratio is 2. Since there are ten terms in this sum, we have $n = 10$. Thus

$$\begin{aligned} S_{10} &= \frac{1}{4} \cdot \frac{1 - 2^{10}}{1 - 2} \\ &= \frac{1}{4} \cdot \frac{1 - 1024}{1 - 2} \\ &= \frac{1}{4} \cdot \frac{-1023}{-1} \\ &= \frac{1023}{4} \end{aligned}$$

EXERCISES 11.4

Identify each of the following as an arithmetic or a geometric series (Exercises 1–10).

1. $3 + 7 + 11 + 15 + \cdots$
2. $\frac{1}{2} + \frac{1}{4} + \frac{1}{8} + \frac{1}{16} + \cdots$
3. $2 + (-2) + 2 + (-2) + \cdots$
4. $1 + \frac{1}{2} + 0 + (-\frac{1}{2}) + \cdots$
5. $a + a^2 + a^3 + a^4 + \cdots$
6. $1 + (-\frac{1}{3}) + \frac{1}{9} + (-\frac{1}{27}) + \cdots$
7. $2 + 8 + 32 + \cdots$
8. $\frac{2}{3} + \frac{4}{9} + \frac{8}{27} + \cdots$
9. $(a + 2z) + (2a + 3z) + (3a + 4z) + \cdots$
10. $\sqrt{2} + 2 + 2\sqrt{2} + \cdots$

Find the common ratio and the next three terms of each of the following geometric progressions (Exercises 11–20).

11. $\frac{1}{2}, \frac{1}{4}, \frac{1}{8}, \ldots$
12. $1, 2, 4, 8, \ldots$
13. $3, -6, 12, \ldots$
14. $48, 24, 12, \ldots$
15. $a, \dfrac{1}{a}, \dfrac{1}{a^3}, \ldots$
16. $\frac{3}{2}, \frac{1}{2}, \frac{1}{6}, \ldots$
17. $\dfrac{a}{b}, \dfrac{a^2}{b}, \dfrac{a^3}{b}, \ldots$
18. $\dfrac{1}{x}, x, x^3, \ldots$
19. $\dfrac{a}{b}, -1, \dfrac{b}{a}, \ldots$
20. $\dfrac{x}{2y}, \dfrac{x}{y}, \dfrac{2x}{y}, \ldots$

Find the indicated term of each of the following geometric progressions (Exercises 21–30).

21. $24, 48, 96, \ldots$; 6th term.
22. $-a, a^2, -a^3, \ldots$; 10th term.
23. $\frac{1}{3}, -\frac{1}{9}, \frac{1}{27}, \ldots$; 5th term.
24. $16, -4, 1, \ldots$; 12th term.
25. $\sqrt{3}, \sqrt{6}, 2\sqrt{6}, \ldots$; 5th term.
26. $\dfrac{1}{5^4}, -\dfrac{1}{5^3}, \dfrac{1}{5^2}, \ldots$; 8th term.
27. $100, -10, 1, \ldots$; 14th term.
28. $6, -2, \frac{2}{3}, \ldots$; 6th term.
29. $27, 9, 3, \ldots$; 7th term.
30. $-2, 6, -18, \ldots$; 15th term.

Find each of the following sums (Exercises 31–38).

31. $\sum_{i=1}^{5} 3^i$
32. $\sum_{i=1}^{8} (\frac{1}{2})^i$
33. $\sum_{i=1}^{10} 2^{i-1}$
34. $\sum_{i=3}^{7} \dfrac{1}{2^{i-1}}$
35. $\sum_{i=1}^{5} (\frac{1}{3})^{i-1}$
36. $\sum_{i=1}^{7} 12(\frac{3}{4})^{i-1}$
37. $\sum_{i=1}^{4} (-\frac{1}{2})^{i-1}$
38. $\sum_{i=1}^{3} (\frac{3}{5})^{-i}$

11.7 ◆ Infinite Geometric Series

Let us consider the infinite geometric sequence

$$\tfrac{1}{2}, \tfrac{1}{4}, \tfrac{1}{8}, \tfrac{1}{16}, \ldots$$

We note the sums of finite portions:

$$S_1 = \tfrac{1}{2}$$
$$S_2 = \tfrac{1}{2} + \tfrac{1}{4} = \tfrac{3}{4}$$
$$S_3 = \tfrac{1}{2} + \tfrac{1}{4} + \tfrac{1}{8} = \tfrac{7}{8}$$
$$S_4 = \tfrac{1}{2} + \tfrac{1}{4} + \tfrac{1}{8} + \tfrac{1}{16} = \tfrac{15}{16}$$
$$S_5 = \tfrac{1}{2} + \tfrac{1}{4} + \tfrac{1}{8} + \tfrac{1}{16} + \tfrac{1}{32} = \tfrac{31}{32}$$

and so forth. It is intuitively evident that the sum of the terms of this sequence is less than 1 no matter how many terms we consider. It is also intuitively clear that the sum can be made as close to 1 as we wish by taking sufficiently many terms.

We express this situation by saying that S_n *approaches 1 as a limit as n becomes indefinitely large* or by saying *the limit of S_n is 1 when n becomes larger and larger*. We indicate this by the symbol

$$\lim_{n \to \infty} S_n = 1$$

When this limit exists, it is called the **sum,** denoted by S, of the infinite geometric series.

Not all infinite geometric progressions have a sum. To illustrate this, let us consider the geometric progression

$$2, 4, 8, 16, \ldots, 2^n, \ldots$$

In this case

$$S_1 = 2$$
$$S_2 = 2 + 4 = 6$$
$$S_3 = 2 + 4 + 8 = 14$$
$$S_4 = 2 + 4 + 8 + 16 = 30$$

and so forth. It is obvious that S_n becomes arbitrarily large as n gets larger and larger.

Our task is to determine a condition under which an infinite geometric series has a sum in the above sense and to find a formula for that sum, when it exists. We know that the sum of n terms of a finite geometric sequence is given by

11.7 INFINITE GEOMETRIC SERIES

$$S_n = a\frac{1-r^n}{1-r} \quad (r \neq 1)$$

It can be proved, although we shall not prove it here, that as n becomes larger and larger, r^n becomes closer and closer to 0 if $-1 < r < 1$—that is, $|r| < 1$.

For example, if $r = \frac{1}{2}$, then

$$r^2 = (\tfrac{1}{2})^2 = \tfrac{1}{4}$$
$$r^3 = (\tfrac{1}{2})^3 = \tfrac{1}{8}$$
$$r^4 = (\tfrac{1}{2})^4 = \tfrac{1}{16}$$
$$r^5 = (\tfrac{1}{2})^5 = \tfrac{1}{32}$$

and so forth.

Returning to

$$S_n = a\frac{1-r^n}{1-r}$$

we see that if n is very large, then r^n is very small when $-1 < r < 1$ and $1 - r^n$ is close to 1. Thus S_n approaches $\dfrac{a}{1-r}$. We define the sum of an infinite geometric series

$$a + ar + ar^2 + ar^3 + \cdots + ar^n + \cdots$$

when $|r| < 1$, to be

$$S = \frac{a}{1-r}$$

EXAMPLE 1. Find the sum of the infinite geometric series

$$3 + \tfrac{3}{2} + \tfrac{3}{4} + \tfrac{3}{8} + \cdots$$

SOLUTION. We see by inspection that $r = \frac{1}{2}$. Using the formula above, we have

$$S = \frac{3}{1-\tfrac{1}{2}}$$
$$= \frac{3}{\tfrac{1}{2}}$$
$$= 6$$

EXAMPLE 2. Express the repeating decimal 0.71313131313... as a common fraction.

SOLUTION. The given decimal may be written in the form

$$0.7 + 0.013 + 0.00013 + 0.0000013 + \cdots$$

The terms

$$0.013 + 0.00013 + 0.0000013 + \cdots$$

form an infinite geometric series whose first term is 0.013 and whose common ratio is 0.01. The sum of this geometric series is

$$S = \frac{0.013}{1 - 0.01}$$

$$= \frac{0.013}{0.99}$$

$$= \frac{13}{990}$$

Then

$$\begin{aligned} 0.7 + 0.013 + 0.00013 + 0.0000013 + \cdots &= 0.7 + \frac{13}{990} \\ &= \frac{7}{10} + \frac{13}{990} \\ &= \frac{693}{990} + \frac{13}{990} \\ &= \frac{706}{990} \\ &= \frac{353}{495} \end{aligned}$$

EXERCISES 11.5

Find the sum of each of the following infinite geometric series (Exercises 1–10).

1. $\frac{1}{3} + \frac{1}{9} + \frac{1}{27} + \frac{1}{81} + \cdots$
2. $6 + (-2) + \frac{2}{3} + (-\frac{2}{9}) + \cdots$
3. $\frac{3}{2} + \frac{3}{4} + \frac{3}{8} + \frac{3}{16} + \cdots$
4. $6 + \frac{6}{10} + \frac{6}{100} + \frac{6}{1000} + \cdots$
5. $1 - \frac{1}{4} + \frac{1}{16} - \frac{1}{64} + \cdots$
6. $\frac{1}{4} + \frac{1}{16} + \frac{1}{64} + \frac{1}{256} + \cdots$
7. $100 + 25 + \frac{25}{4} + \frac{25}{16} + \cdots$
8. $48 - 12 + 3 - \frac{3}{4} + \cdots$
9. $-50 + 10 - 2 + \frac{2}{5} - \cdots$
10. $\sqrt{2} + 1 + \dfrac{1}{\sqrt{2}} + \dfrac{1}{2} + \cdots$

Express each of the following decimal numerals as a common fraction (Exercises 11–20).

11. 0.3333...
12. 0.010101...
13. 0.1666...
14. 0.141141141...
15. 4.393939...
16. 0.838383...
17. 3.139139139...
18. 3.232323...
19. 0.315315315...
20. 8.036036036...

11.8 THE BINOMIAL EXPANSION

21. A rubber ball is dropped from a height of 6 feet. Each time it strikes the floor, it rebounds to approximately three-fourths of the height from which it fell. What, approximately, is the total distance traveled by the ball before it comes to rest?
22. In a square of side 12 inches, a second square is inscribed by joining the midpoints of the sides, in order. In the second square, a third square is inscribed by the same method. If this process is continued indefinitely, what is the sum of the perimeters of all the squares?

11.8 ★ The Binomial Expansion

By multiplication, we may verify the following identities:

$$(a + b)^1 = a + b$$
$$(a + b)^2 = a^2 + 2ab + b^2$$
$$(a + b)^3 = a^3 + 3a^2b + 3ab^2 + b^3$$
$$(a + b)^4 = a^4 + 4a^3b + 6a^2b^2 + 4ab^3 + b^4$$
$$(a + b)^5 = a^5 + 5a^4b + 10a^3b^2 + 10a^2b^3 + 5ab^4 + b^5$$

By examining the above results, we wish to determine a general formula for the expansion of $(a + b)^n$. We observe that in all of the above cases:

1. The first term is a^n.
2. In each succeeding term, the exponent of a is one less and that of b one more than in the term immediately preceding, and in every term the sum of the exponents is n.
3. The coefficient of the second term is $n = \dfrac{n}{1}$.
4. The coefficient of the third term is $\dfrac{n(n-1)}{1 \cdot 2}$.
5. The coefficient of the fourth term is $\dfrac{n(n-1)(n-2)}{1 \cdot 2 \cdot 3}$.

The result of these examples can be generalized to obtain the **binomial expansion**:

$$(a + b)^n = a^n + \frac{n}{1}a^{n-1}b + \frac{n(n-1)}{1 \cdot 2}a^{n-2}b^2 + \frac{n(n-1)(n-2)}{1 \cdot 2 \cdot 3}a^{n-3}b^3$$
$$+ \cdots + \frac{n(n-1)(n-2) \cdots (n-r+2)}{1 \cdot 2 \cdot 3 \cdots (r-1)}a^{n-r+1}b^{r-1} + \cdots + b^n$$

where the rth term is

$$\frac{n(n-1)(n-2) \cdots (n-r+2)}{1 \cdot 2 \cdot 3 \cdots (r-1)}a^{n-r+1}b^{r-1}$$

We can use this formula, called the **binomial theorem,** to find the expansion of $(x - 2)^5$. In this case, a is x and b is -2 in the binomial theorem.

$$(x - 2)^5 = x^5 + \frac{5}{1}x^4(-2) + \frac{5 \cdot 4}{1 \cdot 2}x^3(-2)^2 + \frac{5 \cdot 4 \cdot 3}{1 \cdot 2 \cdot 3}x^2(-2)^3$$
$$+ \frac{5 \cdot 4 \cdot 3 \cdot 2}{1 \cdot 2 \cdot 3 \cdot 4}x(-2)^4 + \frac{5 \cdot 4 \cdot 3 \cdot 2 \cdot 1}{1 \cdot 2 \cdot 3 \cdot 4 \cdot 5}(-2)^5$$
$$= x^5 + 5x^4(-2) + 10x^3(-2)^2 + 10x^2(-2)^3 + 5x(-2)^4 + (-2)^5$$
$$= x^5 - 10x^4 + 40x^3 - 80x^2 + 80x - 32$$

We usually use the symbol $n!$ in writing the binomial expansion. We define the symbol $n!$ to mean the product of all positive integers from 1 to n inclusive. Thus

$$n! = (1)(2)(3) \cdots (n)$$

($n!$ is read: "n factorial").

For example,

$$6! = (1)(2)(3)(4)(5)(6) = 720$$
$$5! = (1)(2)(3)(4)(5) = 120$$

Since

$$n! = (1)(2)(3) \cdots (n - 1)(n)$$

and

$$(n - 1)! = (1)(2)(3) \cdots (n - 1)$$

we see that

$$n! = n(n - 1)!$$

Thus

$$6! = (6)(5!) = 6(120) = 720$$
$$5! = (5)(4!) = 5(24) = 120$$

For $n = 1$ we have

$$1! = (1)(0!)$$

Since $1! = 1$ and $1! = (1)(0!)$, we define

$$\mathbf{0! = 1}$$

We may now write the binomial expansion as

$$(a + b)^n = a^n + \frac{n}{1!}a^{n-1}b + \frac{n(n - 1)}{2!}a^{n-2}b^2 + \frac{n(n - 1)(n - 2)}{3!}a^{n-3}b^3$$
$$+ \cdots + \frac{n(n - 1)(n - 2) \cdots (n - r + 2)}{(r - 1)!}a^{n-r+1}b^{r-1} + \cdots + b^n$$

11.8 THE BINOMIAL EXPANSION

Since
$$n = \frac{n!}{(n-1)!}$$
$$n(n-1) = \frac{n!}{(n-2)!}$$
$$n(n-1)(n-2) = \frac{n!}{(n-3)!}$$

and
$$n(n-1)(n-2) \cdots (n-r+2) = \frac{n!}{(n-r+1)!}$$

we may also write the binomial expansion as

$$(a+b)^n = a^n + na^{n-1}b + \frac{n(n-1)}{2!}a^{n-2}b^2 + \frac{n(n-1)(n-2)}{3!}a^{n-3}b^3$$
$$+ \cdots + \frac{n!}{(r-1)!(n-r+1)!}a^{n-r+1}b^{r-1} + \cdots + b^n$$

EXAMPLE 1. Write the first three terms of $\left(x + \dfrac{1}{x}\right)^{10}$ and simplify.

SOLUTION. The first three terms are

$$x^{10} + 10x^9\left(\frac{1}{x}\right) + \frac{10 \cdot 9}{1 \cdot 2}x^8\left(\frac{1}{x}\right)^2 = x^{10} + 10x^8 + 45x^6$$

EXAMPLE 2. Find the fifth term of $(x-2)^{12}$.

SOLUTION. Using the formula for the rth term we have with $r = 5$, $n = 12$, $a = x$, and $b = -2$

$$\frac{12 \cdot 11 \cdot 10 \cdot 9}{1 \cdot 2 \cdot 3 \cdot 4}x^{12-5+1}(-2)^{5-1} = 495x^8(-2)^4$$
$$= 7920x^8$$

EXERCISES 11.6

Write each of the following in expanded form and simplify (Exercises 1–10).

1. $(x+y)^6$
2. $(x+2)^4$
3. $(x-3)^5$
4. $(2x-y)^5$
5. $(a-2b)^7$
6. $(1-x^2)^6$
7. $(y+\sqrt{2})^4$
8. $(y^3+2a)^4$
9. $(\sqrt{x}-\sqrt{y})^5$
10. $(2z+y\sqrt{2})^4$

Write the first four terms of each of the following and simplify (Exercises 11–16).

11. $(a + 3b)^{10}$ 14. $\left(x - \dfrac{2}{x}\right)^{10}$

12. $(x + \tfrac{1}{2})^{12}$ 15. $(2x - 3y)^{10}$

13. $(y - \tfrac{1}{4})^{15}$ 16. $(x + \sqrt{2})^{19}$

Simplify (Exercises 17–24).

17. $\dfrac{8!}{4!}$ 21. $\dfrac{(12!)(6!)}{(8!)(10!)}$

18. $\dfrac{9!}{2!}$ 22. $\dfrac{(4+2)!}{4!}$

19. $\dfrac{(8-2)!}{(3+4)!}$ 23. $\dfrac{(h+2)!}{h!}$

20. $\dfrac{10!}{(3!)(7!)}$ 24. $\dfrac{(n+1)!}{(n-1)!}$

25. Find the seventh term of $(x + y)^7$.
26. Find the fourth term of $(a - b)^9$.
27. Find the fifth term of $(x^2 - 2)^{10}$.
28. Find the eighth term of $\left(x - \dfrac{1}{x}\right)^{12}$.
29. Find the middle term of $\left(x - \dfrac{2}{x}\right)^{12}$.
30. Find the middle term of $\left(y^2 - \dfrac{3}{y}\right)^{20}$.

CHAPTER 11 REVIEW

1. Write the next three terms in each of the following arithmetic sequences.

   (a) $7, 3, -1, \ldots$ (b) $2, -\tfrac{1}{2}, -3, \ldots$
   (c) $1 + \sqrt{2},\ 1 - \sqrt{2},\ 1 - 3\sqrt{2}, \ldots$

2. Write in expanded form.

   (a) $\sum_{i=1}^{5} (2 + i)$ (c) $\sum_{i=1}^{6} i^3$

   (b) $\sum_{i=1}^{8} (i^2 - 3)$ (d) $\sum_{i=1}^{9} 2^i$

CHAPTER 11 REVIEW

3. Given the arithmetic progression: $-6, -1, 4, \ldots$.

   (a) Find the tenth term.
   (b) Find the sum of the first ten terms.

4. Write the next three terms in each of the following geometric progressions.

   (a) $1, \frac{1}{4}, \frac{1}{16}, \ldots$
   (b) $\frac{1}{3}, \frac{1}{9}, \frac{1}{27}, \ldots$
   (c) $2, 4, 8, \ldots$

5. Given the geometric progression: $12, 4, \frac{4}{3}, \ldots$:

   (a) Find the eighth term.
   (b) Find the sum of the first eight terms.

6. Find each of the following sums.

   (a) $\sum_{i=1}^{5} 3^i$ (c) $\sum_{i=1}^{15} (5-i)$
   (b) $\sum_{i=1}^{10} (3i-2)$ (d) $\sum_{i=1}^{10} (\frac{1}{2})^i$

7. Mr. Hagy is offered a position at a salary that starts at $6500 a year and increases yearly by $150. How much would his total earnings amount to if he worked ten years under this salary schedule?

8. Mrs. Hayward bought some land for $8000. Its value increased 50% of the cost each year. Mrs. Hayward sold the land 5 years after she bought it. How much was the land worth when she sold it?

9. Find the sum of the following infinite geometric series.

   (a) $1 + \frac{1}{4} + \frac{1}{16} + \cdots + \frac{1}{4^{n-1}} + \cdots$
   (b) $0.1 + 0.01 + 0.001 + \cdots + (10)^{-n} + \cdots$

10. Express each of the following decimal numerals as a common fraction.

    (a) $0.88888\ldots$
    (b) $36.205205205\ldots$

Write each of the following in expanded form and simplify (Exercises 11–14).

11. $(a+b)^5$ 13. $\left(\frac{x}{3} - \frac{1}{x}\right)^6$

12. $(3x+2y)^4$ 14. $\left(\frac{5}{x} - x\right)^4$

Write the first four terms of each of the following (Exercises 15–18).

15. $(x+y)^{12}$

17. $\left(x - \dfrac{1}{x}\right)^{20}$

16. $(a^2 - 2b^2)^{16}$

18. $(x^2 + 5y)^{18}$

19. Find the eighth term of $\left(x^2 - \dfrac{1}{x}\right)^{15}$.

20. Find the middle term of $\left(x^2 + \dfrac{1}{x^2}\right)^{20}$.

CHAPTER 12

Exponential and Logarithmic Functions

12.1 ♦ The Exponential Function

We have assigned a meaning to the symbol a^x when a is any real number and x is any rational number (except for those cases in which division by zero or taking an even root of a negative number is involved). We now inquire whether we can interpret powers with irrational numbers as exponents to be real numbers, such as

$$a^{\sqrt{2}}, a^{\pi}, a^{-\sqrt{5}}$$

We know that irrational numbers can be approximated by rational numbers to as great a degree of accuracy as desired. That is, $\sqrt{2} \doteq 1.4$, $\sqrt{2} \doteq 1.41$, $\sqrt{2} \doteq 1.414$, and so forth. It can be proved, but we shall accept it without proof, that if $a > 1$ and x and y are rational numbers with $x > y$, then $a^x > a^y$. Since $(1.5)^x$ is defined when x is rational, we have the following sequence of inequalities

$$(1.5)^1 < (1.5)^{\sqrt{2}} < (1.5)^2$$
$$(1.5)^{1.4} < (1.5)^{\sqrt{2}} < (1.5)^{1.5}$$
$$(1.5)^{1.41} < (1.5)^{\sqrt{2}} < (1.5)^{1.42}$$
$$(1.5)^{1.414} < (1.5)^{\sqrt{2}} < (1.5)^{1.415}$$

and so forth, where $(1.5)^{\sqrt{2}}$ is a number lying between the number on the left and the number on the right. We can continue this process indefinitely so that the difference between the number on the left and that on the right is as small as we please. We assume then, that there is just one number $(1.5)^{\sqrt{2}}$ that will satisfy the inequalities no matter how long we continue this process.

It can also be shown that if $0 < a < 1$, a^x denotes a definite number for every real value of x—irrational as well as rational.

EXPONENTIAL AND LOGARITHMIC FUNCTIONS

Without further discussion, let us assume that for each real x there is one and only one number a^x, for $a > 0$.

Since for each real x there is one and only one number a^x, where $a > 0$, the equation

$$y = a^x \ (a > 0)$$

specifies a function. If a is 1, we have $1^x = 1$, a **constant function**, whose graph is a line parallel to the x-axis. If $0 < a < 1$ or $a > 1$, then the equation $y = a^x$ specifies an **exponential function**.

Let us graph the exponential function specified by

$$y = (1.5)^x$$

Some ordered pairs that belong to this function are

$$(-4, 0.2), (-3, 0.3), (-2, 0.44), (-1, 0.67), (0, 1),$$
$$(1, 1.5), (2, 2.25), (3, 3.38), (4, 5.06)$$

(some of the y-coordinates are approximations). Plotting these points and connecting them with a smooth curve, we have the graph in Figure 12.1.

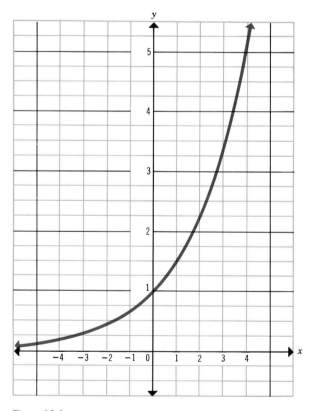

Figure 12.1

12.1 THE EXPONENTIAL FUNCTION

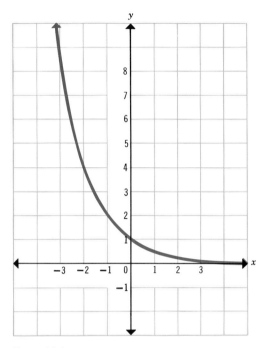

Figure 12.2

Now let us graph the exponential function specified by

$$y = (\tfrac{1}{2})^x$$

Some ordered pairs of this function are

$$(-3, 8), (-2, 4), (-1, 2), (0, 1), (1, \tfrac{1}{2}), (2, \tfrac{1}{4}), (3, \tfrac{1}{8})$$

The graph of this function is shown in Figure 12.2.

Notice that the graph of $y = (1.5)^x$ goes up as x increases and the graph of $y = (\tfrac{1}{2})^x$ goes down as x increases. In general, if $a > 1$, the exponential function increases as x increases and has the general shape of the graph in Figure 12.1. If $0 < a < 1$, the function decreases as x increases and has the general shape of the graph in Figure 12.2.

EXERCISES 12.1

Draw the graphs of the exponential functions specified by the equations below (Exercises 1–10).

1. $y = 2^x$
2. $y = 3^x$
3. $y = 4^{-x}$
4. $y = (\tfrac{1}{4})^x$
5. $y = 2^{2x}$
6. $y = 3^{-x}$

7. $y = (\frac{1}{3})^{-x}$ 9. $y = 10^x$
8. $y = (\frac{1}{10})^x$ 10. $y = 4^{2x}$

11. Tell whether the exponential functions given by the following equations have graphs that increase as x increases or that decrease as x increases.
    (a) $y = 3^x$ (d) $y = 4^{-x}$
    (b) $y = (1.6)^x$ (e) $y = (\frac{1}{10})^{2x}$
    (c) $y = (\frac{2}{3})^x$ (f) $y = 2(1.5)^x$
12. Using the laws of exponents, show that $(\frac{1}{2})^x$ and 2^{-x} are identical.
13. Using laws of exponents, show that 2^{3x} may be written 8^x.
14. Using laws of exponents, show that 2^{x-3} may be written $\frac{1}{8}(2^x)$.

12.2 ◆ The Logarithmic Function

If a is a positive real number other than 1, x is a real number greater than 0, and

$$a^y = x$$

we call y the **logarithm of x to the base a**. We use the symbol

$$\log_a x = y$$

to indicate the logarithm of x to the base a. Thus,

since $2^3 = 8$, $\log_2 8 = 3$

since $10^2 = 100$, $\log_{10} 100 = 2$

since $2^{-2} = \frac{1}{4}$, $\log_2 (\frac{1}{4}) = -2$

It should be observed that $a^y = x$ and $y = \log_a x$ are equivalent equations, both specifying the same function. Thus exponential equations may be written in **logarithmic form** and logarithmic equations may be written in **exponential form**. For example:

Exponential Form	Logarithmic Form
$3^2 = 9$	$\log_3 9 = 2$
$5^3 = 125$	$\log_5 125 = 3$
$(\frac{1}{2})^{-1} = 2$	$\log_{1/2} 2 = -1$

The function specified by

$$y = \log_a x$$

is called a **logarithmic function**. The graph of the logarithmic function $y = \log_{10} x$ is shown in Figure 12.3.

12.2 THE LOGARITHMIC FUNCTION

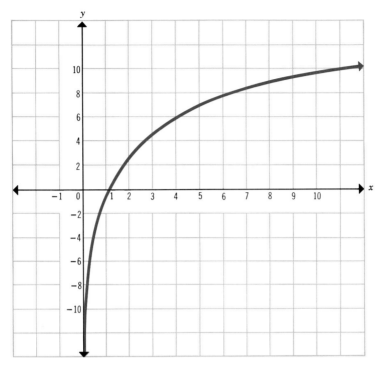

Figure 12.3

EXERCISES 12.2

Express in logarithmic form (Exercises 1–12).

1. $2^2 = 4$
2. $5^2 = 25$
3. $10^2 = 100$
4. $(\frac{1}{2})^3 = \frac{1}{8}$
5. $2^{-1} = \frac{1}{2}$
6. $3^0 = 1$
7. $100^{1/2} = 10$
8. $(\frac{1}{4})^{-2} = 16$
9. $10^{-2} = 0.01$
10. $16^{-1/2} = \frac{1}{4}$
11. $10^3 = 1000$
12. $5^{-2} = \frac{1}{25}$

Express in exponential notation (Exercises 13–20).

13. $\log_2 64 = 6$
14. $\log_5 25 = 2$
15. $\log_4 1 = 0$
16. $\log_{10} 0.01 = -2$
17. $\log_{1/2} 8 = -3$
18. $\log_{10} 1000 = 3$
19. $\log_8 4 = \frac{2}{3}$
20. $\log_{1/5} 125 = -3$

Find the value of each of the following (Exercises 21–28).

21. $\log_{10} 100$
22. $\log_2 16$
23. $\log_3 27$
24. $\log_{10} 1$
25. $\log_{10} 0.01$
26. $\log_7 49$
27. $\log_2 2$
28. $\log_{10} 10$

Graph the functions specified by the following equations (Exercises 29–30).

29. $y = \log_{10} x$ 30. $y = \log_2 x$

12.3 ◆ Laws of Logarithms

By definition, a logarithm is an exponent, hence, if $a > 0$ and x_1 and x_2 are positive, then:

LAW 1 $$\log_a (x_1 \cdot x_2) = \log_a x_1 + \log_a x_2$$

The validity of this law can be seen from the laws of exponents. Let

$$x_1 = a^M \quad \text{and} \quad x_2 = a^N$$

Then

$$x_1 \cdot x_2 = a^M \cdot a^N = a^{M+N}$$

But, by definition of a logarithm,

$$\log_a (x_1 \cdot x_2) = M + N$$

From the definition of logarithm, $M = \log_a x_1$ and $N = \log_a x_1$, since $x_1 = a^M$ and $x_2 = a^N$, so

$$\log_a (x_1 \cdot x_2) = \log_a x_1 + \log_a x_2$$

The second law of logarithms states that if $a > 0$ and x_1 and x_2 are positive, then:

LAW 2 $$\log_a \frac{x_1}{x_2} = \log_a x_1 - \log_a x_2$$

We can establish the validity of Law 2 in the same way as we did Law 1. Let $x_1 = a^M$ and $x_2 = a^N$. Then

$$\frac{x_1}{x_2} = \frac{a^M}{a^N} = a^{M-N}$$

By the definition of logarithm,

$$\log_a \frac{x_1}{x_2} = M - N$$

But since $x_1 = a^M$ and $x_2 = a^N$, $\log_a x_1 = M$ and $\log_a x_2 = N$, so that

$$\log_a \frac{x_1}{x_2} = \log_a x_1 - \log_a x_2$$

12.3 LAWS OF LOGARITHMS

Another law of logarithms, which we shall accept without proof, is that for $a > 0$ and $x > 0$, and m any real number:

LAW 3 $$\log_a x^m = m \log_a x$$

Before illustrating the use of these laws, we state three properties of the logarithmic function, which we shall also accept without proof. For $a > 0$:

1. If $M = N$, then $\log_a M = \log_a N$.
2. If $\log_a M = \log_a N$, then $M = N$.
3. If $M = N$, then $a^M = a^N$.

EXAMPLE 1. Express $\frac{1}{4} \log_a x - \frac{3}{4} \log_a y$ as a single logarithm with coefficient 1.

SOLUTION. By the third law of logarithms,

$$\tfrac{1}{4} \log_a x - \tfrac{3}{4} \log_a y = \log_a x^{1/4} - \log_a y^{3/4}$$

By the second law of logarithms,

$$\log_a x^{1/4} - \log_a y^{3/4} = \log_a \frac{x^{1/4}}{y^{3/4}}$$

EXAMPLE 2. Given $\log_{10} 2 = 0.3010$ and $\log_{10} 7 = 0.8451$, find $\log_{10} \tfrac{7}{8}$.

SOLUTION. By the second law of logarithms,

$$\log_{10} \tfrac{7}{8} = \log_{10} 7 - \log_{10} 8$$

Since $8 = 2^3$, the third law of logarithms gives

$$\begin{aligned}
\log_{10} \tfrac{7}{8} &= \log_{10} 7 - 3 \log_{10} 2 \\
&= 0.8451 - 3(0.3010) \\
&= 0.8451 - 0.9030 \\
&= -0.0579
\end{aligned}$$

EXERCISES 12.3

Express as a single logarithm with coefficient 1 (Exercises 1–10).

1. $\log_a x + \log_a y$
2. $\log_a x - \log_a y$
3. $3 \log_a x$
4. $3 \log_a x + 2 \log_a y$
5. $\log_a x + 2 \log_a y$
6. $\frac{1}{2} \log_a x - 3 \log_a y$
7. $\frac{1}{4}(\log_a x + \log_a y)$
8. $\log_a (x - y) + 3 \log_a x$
9. $3 \log_a x - \frac{4}{3} \log_a y$
10. $-\frac{1}{2} \log_a x$

Given that $\log_{10} 2 = 0.301$, $\log_{10} 3 = 0.477$, and $\log_{10} 7 = 0.845$, find the following (Exercises 11–20).

11. $\log_{10} 6$
12. $\log_{10} 21$
13. $\log_{10} 9$
14. $\log_{10} 36$
15. $\log_{10} \frac{7}{16}$
16. $\log_{10} \sqrt{2}$
17. $\log_{10} \sqrt[3]{\frac{14}{3}}$
18. $\log_{10} 63$
19. $\log_{10} \frac{3}{2}$
20. $\log_{10} \sqrt[3]{\frac{27}{2}}$

12.4 ◆ Common Logarithms

There are two logarithmic functions of importance in mathematics. The one we shall study is the **common logarithmic function,** which is specified by the equation

$$y = \log_{10} x$$

Values of this function are called **logarithms to the base ten** or **common logarithms.** By definition of a common logarithm, we can readily determine

$$\log_{10} 10 = 1, \text{ since } 10^1 = 10$$
$$\log_{10} 100 = 2, \text{ since } 10^2 = 100$$
$$\log_{10} 1000 = 3, \text{ since } 10^3 = 1000$$

and so forth. Also

$$\log_{10} 1 = 0, \text{ since } 10^0 = 1$$
$$\log_{10} 0.1 = -1, \text{ since } 10^{-1} = 0.1$$
$$\log_{10} 0.01 = -2, \text{ since } 10^{-2} = 0.01$$
$$\log_{10} 0.001 = -3, \text{ since } 10^{-3} = 0.001$$

and so forth.

To determine the common logarithm, or $\log_{10}$, of a positive number which is not an integral power of 10, we use a table called a **log table.** We can find an approximation to the common logarithm of that number in the table. Figure 12.4 shows a portion of such a table. In this log table all the decimal points have been omitted. In the column headed "N," a decimal point should occur after the first digit. In the other columns, a decimal point should occur before the first digit. The table in Figure 12.4 is repeated below, with decimal points in their proper places.

Figure 12.5 shows how to use the table to determine that $\log_{10} 4.36 = 0.6395$. First, find "43" (the first two digits in 4.36) in the column labeled N, and then find "6" (the third digit in 4.36) in the top row. Now go to the right from "43" and down from "6" until you find "6395." This means

12.4 COMMON LOGARITHMS

N	0	1	2	3	4	5	6	7	8	9
39	5911	5922	5933	5944	5955	5966	5977	5988	5999	6010
40	6021	6031	6042	6053	6064	6075	6085	6096	6107	6117
41	6128	6138	6149	6160	6170	6180	6191	6201	6212	6222
42	6232	6243	6253	6263	6274	6284	6294	6304	6314	6325
43	6335	6345	6355	6365	6375	6385	6395	6405	6415	6425
44	6435	6444	6454	6464	6474	6484	6493	6503	6513	6522

Figure 12.4

$\log_{10} 4.36 = 0.6395$. Figure 12.5 also shows how to determine that $\log_{10} 3.93 = 0.5944$.

In writing that $\log_{10} 4.36 = 0.6395$, we are really writing only a rational approximation of the logarithm because the logarithms of numbers other than integer powers of ten are irrational and cannot be represented exactly by decimals. It is customary, however, to use an equal sign, and we shall follow the practice.

The table of common logarithms in the Appendix is used to find $\log_{10} x$ where $1 \leq x < 10$. If we wish to find $\log_{10} x$ for values of x outside the range of that table—that is, for $0 < x < 1$ or $x \geq 10$, we first represent, in **scientific notation,** the number whose logarithm we seek, and then we use the laws of logarithms. This means writing x as the product of a number between 1 and 10 and an integral power of 10 (see Chapter 2, Section 2.7). For example,

$$\log_{10} 38.4 = \log_{10} [3.84(10)^1]$$
$$= \log_{10} 3.84 + \log_{10} 10^1$$
$$= 0.5843 + 1$$
$$= 1.5843$$
$$\log_{10} 384 = \log_{10} [3.84(10)^2]$$
$$= \log_{10} 3.84 + \log_{10} 10^2$$
$$= 0.5843 + 2$$
$$= 2.5843$$

N	0	1	2	3	4	5	6	7	8	9
3.9	.5911	.5922	.5933	.5944	.5955	.5966	.5977	.5988	.5999	.6010
4.0	.6021	.6031	.6042	.6053	.6064	.6075	.6085	.6096	.6107	.6117
4.1	.6128	.6138	.6149	.6160	.6170	.6180	.6191	.6201	.6212	.6222
4.2	.6232	.6243	.6253	.6263	.6274	.6284	.6294	.6304	.6314	.6325
4.3	.6335	.6345	.6355	.6365	.6375	.6385	.6395	.6405	.6415	.6522
4.4	.6435	.6444	.6454	.6464	.6474	.6484	.6493	.6503	.6513	.6522

Figure 12.5

250   EXPONENTIAL AND LOGARITHMIC FUNCTIONS

$$\begin{aligned}\log_{10} 3840 &= \log_{10}[3.84(10)^3] \\ &= \log_{10} 3.84 + \log_{10} 10^3 \\ &= 0.5843 + 3 \\ &= 3.5843 \\ \log_{10} 0.0384 &= \log_{10}[3.84(10)^{-2}] \\ &= \log_{10} 3.84 + \log_{10} 10^{-2} \\ &= 0.5843 + (-2) \\ \log_{10} 0.00384 &= \log_{10}[3.84(10)^{-3}] \\ &= \log_{10} 3.84 + \log_{10} 10^{-3} \\ &= 0.5843 + (-3)\end{aligned}$$

Notice that in all the examples above, the result is in the form $c + m$, where c is an integer and $0 < m < 1$; c is called the **characteristic** of the logarithm and m is called the **mantissa**. (The log table is, in fact, a table of mantissas.) We observe that

$$\begin{aligned}\log_{10} 0.00384 &= 0.5843 + (-3) \\ &= 0.5843 + (7 - 10) \\ &= 7.5843 - 10\end{aligned}$$

Both $0.5843 + (-3)$ and $7.5843 - 10$ are valid ways of writing $\log_{10} 0.00384$, but $7.5843 - 10$ is customary in most cases.

It is possible to find a number when its logarithm is given. That is, if $\log_{10} x$ is known, then it is possible to find x. In this case x is called the **antilogarithm** (written: antilog$_{10}$) of $\log_{10} x$. For example, to find the number x if $\log_{10} x = 4.6085$, we proceed as follows:

$$\begin{aligned}\log_{10} x &= 4 + 0.6085 \\ &= \log_{10} 10^4 + 0.6085\end{aligned}$$

The number antilog$_{10}$ 0.6085 can be found by locating the mantissa 0.6085 in the body of the logarithm table and observing that the associated number is 4.06. Hence

$$\begin{aligned}\log_{10} x &= \log_{10} 10^4 + \log_{10} 4.06 \\ &= \log_{10}[4.06(10)^4] \\ &= \log_{10} 40600\end{aligned}$$

Thus $x = 40{,}600$, and we write

$$\text{antilog}_{10} 4.6085 = 40{,}600$$

Thus, to solve an equation such as

$$\log_{10} x = 9.6464 - 10$$

12.4 COMMON LOGARITHMS

we proceed as follows:

$$\log_{10} x = 9.6464 - 10$$
$$= 0.6464 + (-1)$$
$$= 0.6464 + \log_{10} 10^{-1}$$

We find antilog$_{10}$ 0.6464 by locating the mantissa 0.6464 in the table and observing that

$$\text{antilog}_{10}\ 0.6464 = 4.43$$

Thus

$$\log_{10} x = \log_{10} 4.43 + \log_{10} 10^{-1}$$
$$= \log_{10} [4.43(10)^{-1}]$$
$$= \log_{10} 0.443$$

Thus $x = 0.443$.

EXERCISES 12.4

Find each logarithm (Exercises 1–45).

1. $\log_{10} 4.24$
2. $\log_{10} 1.02$
3. $\log_{10} 7.18$
4. $\log_{10} 9.99$
5. $\log_{10} 1.00$
6. $\log_{10} 8.46$
7. $\log_{10} 2.15$
8. $\log_{10} 5.69$
9. $\log_{10} 3.31$
10. $\log_{10} 6.43$
11. $\log_{10} 87.4$
12. $\log_{10} 49.3$
13. $\log_{10} 111$
14. $\log_{10} 69.7$
15. $\log_{10} 232$
16. $\log_{10} 0.444$
17. $\log_{10} 0.505$
18. $\log_{10} 8$
19. $\log_{10} 1.69$
20. $\log_{10} 0.00111$
21. $\log_{10} 0.00707$
22. $\log_{10} 69$
23. $\log_{10} 237,000$
24. $\log_{10} 0.00319$
25. $\log_{10} 0.00427$
26. $\log_{10} 0.00143$
27. $\log_{10} 986,000$
28. $\log_{10} 0.0000927$
29. $\log_{10} 93,100,000$
30. $\log_{10} 2,780$
31. $\log_{10} 0.319$
32. $\log_{10} 87.9$
33. $\log_{10} 0.000495$
34. $\log_{10} 608,000$
35. $\log_{10} 73.8$
36. $\log_{10} 639,000,000$
37. $\log_{10} 0.08$
38. $\log_{10} 0.0032$
39. $\log_{10} 0.00305$
40. $\log_{10} [2.76(10^2)]$
41. $\log_{10} [1.09(10^7)]$
42. $\log_{10} [2.31(10^{-4})]$
43. $\log_{10} [4.16(10^{-2})]$
44. $\log_{10} [1.92(10^3)]$
45. $\log_{10} [2.76(10^{-8})]$

Find each antilogarithm (Exercises 46–75).

46. antilog$_{10}$ 0.9042
47. antilog$_{10}$ 0.9754
48. antilog$_{10}$ 0.6884
49. antilog$_{10}$ 0.9736
50. antilog$_{10}$ 0.9542
51. antilog$_{10}$ 0.0043
52. antilog$_{10}$ 1.4800
53. antilog$_{10}$ 1.8998
54. antilog$_{10}$ 2.6522
55. antilog$_{10}$ 4.7356
56. antilog$_{10}$ 5.8274
57. antilog$_{10}$ 2.8082
58. antilog$_{10}$ 6.7803
59. antilog$_{10}$ 9.4871 − 10
60. antilog$_{10}$ 7.3010 − 10

252 EXPONENTIAL AND LOGARITHMIC FUNCTIONS

61. antilog$_{10}$ 8.6042 − 10
62. antilog$_{10}$ 7.5340 − 10
63. antilog$_{10}$ 5.3181 − 10
64. antilog$_{10}$ 6.3181 − 10
65. antilog$_{10}$ 9.3962 − 10

66. antilog$_{10}$ 8.6484 − 10
67. antilog$_{10}$ 9.7316 − 10
68. antilog$_{10}$ 6.8698 − 10
69. antilog$_{10}$ 4.6767 − 10
70. antilog$_{10}$ 4.1038

71. antilog$_{10}$ 5.1644
72. antilog$_{10}$ 7.2945 − 10
73. antilog$_{10}$ 8.6646 − 10
74. antilog$_{10}$ 5.1644 − 10
75. antilog$_{10}$ 6.6170

12.5 ◆ Linear Interpolation

A function is a set of ordered pairs. A table of logarithms represents just such a set of ordered pairs. For each number, x, there is associated the number $\log_{10} x$, and we have a set of ordered pairs of the form $(x, \log_{10} x)$ displayed in a convenient form. The four-place table we use only gives three digits for the number x and four decimal places for the number $\log_{10} x$. By means of a process called **linear interpolation** this table can be used to find approximations of logarithms for numbers named by four digit numerals.

Suppose it is required to find $\log_{10} 1.414$. A portion of the graph of $y = \log_{10} x$ is shown in Figure 12.6. The curvature is exaggerated to illustrate the principle involved. We use the line segment connecting P_1 and P_2 as an approximation of the curve passing through $P_1 P_2$. If we had a sufficiently large graph of $y = \log_{10} x$, we could find the value of the ordinate (SP) of the point on the curve for which the abscissa is 1.414. Since we cannot do this with table values only, we shall use the value of the ordinate (ST) to the straight line as an approximation to the ordinate of the point on the curve.

Observing Figure 12.6, we see that the triangles $P_1 TR$ and $P_1 P_2 M$ are similar, since $P_2 M$ and TR are perpendicular to $P_1 M$ so that we have two right triangles with a common angle (the angle $TP_1 R$). The sides of similar

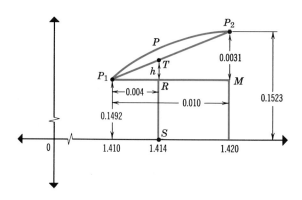

Figure 12.6

triangles are proportional so that

$$\frac{P_1R}{P_1M} = \frac{TR}{P_2M}$$

Since we know three of these four numbers, we can find the fourth. Thus

$$\frac{P_1R}{P_1M} = \frac{1.414 - 1.410}{1.420 - 1.410} = \frac{0.004}{0.010} = \frac{4}{10}$$

Hence

$$\frac{TR}{P_2M} = \frac{4}{10}$$

Since $P_2M = 0.1523 - 0.1492 = 0.0031$, if we take $TR = h$ we have

$$\frac{h}{0.0031} = \frac{4}{10}$$

$$h = \frac{4}{10}(0.0031) = 0.00124 \doteq 0.0012$$

Adding 0.0012 to 0.1492 we see that

$$\log_{10} 1.414 = 0.1504$$

The example below shows a shortened method for the procedure illustrated above.

EXAMPLE 1. Find $\log_{10} 56.24$.

SOLUTION

$$0.10 \left\{ 0.04 \begin{cases} \log_{10} 56.20 = 1.7497 \\ \log_{10} 56.24 = ? \\ \log_{10} 56.30 = 1.7505 \end{cases} h \right\} 0.0008$$

$$\frac{4}{10} = \frac{h}{0.0008}$$

$$h = \frac{4}{10}(0.0008) = 0.00032$$

We round off to four decimal places, taking

$$h = 0.0003$$

Thus

$$\log_{10} 56.24 = 1.7497 + 0.0003 = 1.7500$$

The example below shows the procedure used in finding antilog_{10} of a number whose mantissa is not in the table. Again we are using the principle of linear interpolation.

EXAMPLE 2. Find $\text{antilog}_{10}\ 0.4327$.

SOLUTION

$$0.0016 \left\{ 0.0013 \left\{ \begin{array}{l} \text{antilog}_{10}\ 0.4314 = 2.700 \\ \text{antilog}_{10}\ 0.4327 = \ ? \\ \text{antilog}_{10}\ 0.4330 = 2.710 \end{array} \right\} h \right\} 0.010$$

$$\frac{13}{16} = \frac{h}{0.01}$$

$$h = \frac{13}{16}(0.01)$$

$$= 0.008 \text{ (rounded to three decimal places)}$$

$$\text{Antilog}_{10}\ 0.4327 = 2.700 + 0.008 = 2.708$$

EXERCISES 12.5

Find each logarithm (Exercises 1–20).

1. $\log_{10} 1.432$
2. $\log_{10} 1.876$
3. $\log_{10} 3.107$
4. $\log_{10} 37.26$
5. $\log_{10} 186.9$
6. $\log_{10} 72.96$
7. $\log_{10} 3854$
8. $\log_{10} 0.001342$
9. $\log_{10} 0.07961$
10. $\log_{10} 873,100$
11. $\log_{10} 0.007969$
12. $\log_{10} 0.00008649$
13. $\log_{10} 78.97$
14. $\log_{10} 84.48$
15. $\log_{10} 1387$
16. $\log_{10} 278,300$
17. $\log_{10} 0.001111$
18. $\log_{10} 8.358$
19. $\log_{10} 0.002685$
20. $\log_{10} 93,640,000$

Find each antilogarithm (Exercises 21–40).

21. $\text{antilog}_{10}\ 0.3467$
22. $\text{antilog}_{10}\ 0.1567$
23. $\text{antilog}_{10}\ 1.4689$
24. $\text{antilog}_{10}\ 1.4056$
25. $\text{antilog}_{10}\ 2.1084$
26. $\text{antilog}_{10}\ 3.2963$
27. $\text{antilog}_{10}\ 4.9308$
28. $\text{antilog}_{10}\ 4.0952$
29. $\text{antilog}_{10}\ 5.5589$
30. $\text{antilog}_{10}\ 2.1573$
31. $\text{antilog}_{10}\ (9.1379 - 10)$
32. $\text{antilog}_{10}\ (9.0045 - 10)$
33. $\text{antilog}_{10}\ (8.1456 - 10)$
34. $\text{antilog}_{10}\ (9.4444 - 10)$
35. $\text{antilog}_{10}\ (7.2165 - 10)$
36. $\text{antilog}_{10}\ (8.0314 - 10)$
37. $\text{antilog}_{10}\ (7.0031 - 10)$
38. $\text{antilog}_{10}\ (9.3721 - 10)$
39. $\text{antilog}_{10}\ (7.1134 - 10)$
40. $\text{antilog}_{10}\ (8.3690 - 10)$

12.6 ◆ Logarithmic Computation

Numerical computation may be simplified by using the laws of logarithms. For convenience, we restate the laws of logarithms for logarithms with base 10.

LAW 1: $\qquad \log_{10}(x_1 \cdot x_2) = \log_{10} x_1 + \log_{10} x_2$

LAW 2: $\qquad \log_{10} \dfrac{x_1}{x_2} = \log_{10} x_1 - \log_{10} x_2$

LAW 3: $\qquad \log_{10} x^n = n \log_{10} x$

We also restate the properties of the logarithmic and exponential functions given in Section 12.3.

1. If $M = N$, then $\log_{10} M = \log_{10} N$.
2. If $\log_{10} M = \log_{10} N$, then $M = N$.
3. If $M = N$, then $10^M = 10^N$.

EXAMPLE 1. Compute $(38.6)(42.7)$ by means of logarithms.

SOLUTION. Let $N = (38.6)(42.7)$. Then

$$\begin{aligned}
\log_{10} N &= \log_{10} 38.6 + \log_{10} 42.7 \qquad \text{by Law 1} \\
&= 1.5866 + 1.6304 \\
&= 3.2170 \\
N &= \text{antilog}_{10}\, 3.2170 \\
&= 1648
\end{aligned}$$

EXAMPLE 2. Compute $\dfrac{(7.34)(87.6)}{0.539}$ using logarithms.

SOLUTION. Let $N = \dfrac{(7.34)(87.6)}{0.539}$. By Law 1 and Law 2, we have

$$\begin{aligned}
\log_{10} N &= \log_{10} 7.34 + \log_{10} 87.6 - \log_{10} 0.539 \\
&= 0.8657 + 1.9425 - (9.7316 - 10) \\
&= 3.0766 \\
N &= \text{antilog}_{10}\, 3.0766 \\
&= 1{,}193
\end{aligned}$$

EXAMPLE 3. Compute $\dfrac{\sqrt[5]{-992.7}}{\sqrt[3]{0.2355}}$ using logarithms.

SOLUTION: Since the fifth root of a negative number is negative and the quotient of a negative number divided by a positive number is negative, the required result is negative. In doing the computation we disregard the sign and let

$$N = \frac{\sqrt[5]{992.7}}{\sqrt[3]{0.2358}}$$

Then

$\log_{10} N = \tfrac{1}{5} \log_{10} 992.7 - \tfrac{1}{3} \log_{10} 0.2358$
$\quad\quad\quad = \tfrac{1}{5}(2.9968) - \tfrac{1}{3}(9.3725 - 10)$
$\quad\quad\quad = \tfrac{1}{5}(2.9968) - \tfrac{1}{3}(29.3725 - 30)$ (since we need a characteristic that is divisible by 3)
$\quad\quad\quad = 0.5994 - (9.7908 - 10)$
$\quad\quad\quad = (10.5994 - 10) - (9.7908 - 10)$
$\quad\quad\quad = 0.8086$
$N = \text{antilog}_{10}\, 0.8086 = 6.436$

EXERCISES 12.6

Compute using logarithms.

1. $(8.12)(6.34)$
2. $(1.27)(9.65)$
3. $\dfrac{7.14}{1.34}$
4. $\dfrac{38.9}{4.62}$
5. $(1.73)^3$
6. $\sqrt{4.889}$
7. $\dfrac{(27.8)(19.5)}{(3.82)^4}$
8. $\sqrt{\dfrac{6.93}{2.17}}$
9. $\dfrac{(21.2)\sqrt{8.32}}{(1.76)^2}$
10. $\sqrt[3]{19.7}$
11. $\dfrac{\sqrt{42.6}}{(18.3)(11.9)}$
12. $\dfrac{(17.8)^2(39.4)}{\sqrt{91.2}}$
13. $(2.13)^2(1.24)^3$
14. $\sqrt[5]{13.4}$
15. $\sqrt[3]{(1.86)(2.13)}$
16. $\dfrac{\sqrt{182}}{(1.24)(4.36)}$
17. $\dfrac{\sqrt[3]{1.84}}{(9.16)(82.4)}$
18. $\sqrt{\dfrac{1.09}{0.143}}$
19. $\sqrt[8]{9.09}$
20. $\dfrac{1.86}{59.3}$
21. $\sqrt[3]{(86.4)(119)}$
22. $(1.3)^3(28.1)^2$
23. $(88.9)(4.32)(\sqrt{8.71})$
24. $\dfrac{(7.38)(\sqrt{82})}{(12.3)^3}$
25. $\sqrt[5]{72.9}$
26. $\sqrt{(1.06)(3.97)}$
27. $\dfrac{\sqrt{92.3}\,\sqrt[3]{82}}{(1.7)^2}$

28. $\dfrac{\sqrt{329}}{(68.4)(9.27)}$

29. $(1.86)^{2.4}$

30. $\dfrac{(1.73)(11.4)^3}{\sqrt{96.2}}$

31. $\dfrac{(8.78)^3}{\sqrt[5]{9.73}}$

32. $\dfrac{(\sqrt[4]{87.6})(3.42)}{\sqrt[5]{36.9}}$

33. $\dfrac{\sqrt[3]{1.09}}{\sqrt[4]{0.026}}$

34. $(0.00463)^3$

35. $(0.164)(0.0023)^3$

36. $\dfrac{\sqrt[4]{0.00463}}{(0.0912)(0.0031)}$

37. $\sqrt[5]{\dfrac{0.00364}{(0.092)(3.4)}}$

38. $\sqrt[3]{\dfrac{1.23}{4.79}}\sqrt[5]{\dfrac{18.2}{0.016}}$

39. $\dfrac{(-7.32)^4(0.00964)^3}{-\sqrt{0.181}}$

40. $\dfrac{(-2.73)^3\sqrt{0.09}}{(1.62)(0.00463)^4}$

41. If P dollars are invested at 6% compounded annually, the amount at the end of n years is given by the formula $S = P(1.06)^n$. If $6000 is invested for 12 years, find the amount at the end of 12 years.

42. The time, t, in seconds for one complete oscillation of a pendulum whose length is l feet is given by

$$t = 2\pi\sqrt{\dfrac{l}{g}}$$

where $g = 32.16$ feet per second per second. Find t for a pendulum 2.86 feet long (use $\pi = 3.14$).

43. The volume of a sphere is given by $V = \tfrac{4}{3}\pi r^3$. Find the volume of a sphere whose radius is 5.25 inches (use $\pi = 3.14$).

12.7 ♦ Logarithmic and Exponential Equations

An equation in which the variable appears in an exponent is called an **exponential equation.** The solution sets of some exponential equations may be found by using logarithms.

EXAMPLE 1. Find the solution set of $6^x = 9$.

SOLUTION. Since 6^x is positive for all x, the given equation is equivalent to

$$\log_{10} 6^x = \log_{10} 9$$
$$x \log_{10} 6 = \log_{10} 9$$
$$x = \dfrac{\log_{10} 9}{\log_{10} 6}$$
$$x = \dfrac{0.9542}{0.7782}$$
$$x = 1.2262$$

The solution set is $\{1.2262\}$.

EXAMPLE 2. Find the solution set of $9^{x-1} = 7^x$.

SOLUTION. Since 9^{x-1} and 7^x are positive for all x, the given equation is equivalent to

$$\log_{10} 9^{x-1} = \log_{10} 7^x$$
$$(x - 1) \log_{10} 9 = x \log_{10} 7$$
$$x \log_{10} 9 - \log_{10} 9 = x \log_{10} 7$$
$$x(\log_{10} 9 - \log_{10} 7) = \log_{10} 9$$
$$x = \frac{\log_{10} 9}{\log_{10} 9 - \log_{10} 7}$$
$$x = \frac{0.9542}{0.9542 - 0.8451}$$
$$x = 8.7461$$

The solution set is $\{8.7461\}$.

The term **logarithmic equation** is usually applied to an equation in which the logarithm of a variable occurs. Solution sets of some logarithmic equations can be found.

EXAMPLE 3. Find the solution set of $\log_{10}(x + 1) - \log_{10} 2 = \log_{10} 8$.

SOLUTION. We have the equivalent equations:

$$\log_{10}(x + 1) = \log_{10} 8 + \log_{10} 2$$
$$\log_{10}(x + 1) = \log_{10}[(8)(2)]$$
$$\log_{10}(x + 1) = \log_{10} 16$$
$$x + 1 = 16$$
$$x = 15$$

The solution set is $\{15\}$.

EXAMPLE 4. Find the solution set of $\log_{10}(3x + 2) = 3$.

SOLUTION. The given equation is equivalent to

$$\log_{10}(3x + 2) = \log_{10} 10^3$$
$$3x + 2 = 1000$$
$$3x = 998$$
$$x = \tfrac{998}{3}$$

The solution set is $\{\tfrac{998}{3}\}$.

EXERCISES 12.7

Find the solution sets.

1. $4^x = 5$
2. $3^x = 16$
3. $2^x = 24$
4. $3^x = 2^{x+2}$
5. $1000^x = 68$
6. $5^x = 32$
7. $3^{2x-1} = 78$
8. $2^{-x} = 27$
9. $10^x = 50^{x+2}$
10. $2^{x-2} = 24$
11. $\log_{10} x + \log_{10} 4 = 5$
12. $\log_{10} x + \log_{10} (x + 2) = 3$
13. $\log_{10} (x + 1) - \log_{10} 4 = \log_{10} 16$
14. $\log_{10} (2x + 3) = 7$
15. $\log_{10} (2x + 7) = \log_{10} 4 + \log_{10} (2 - x)$
16. $\log_{10} x^5 = 4$
17. $\log_{10} (2x + 5) = 2$
18. $x \log_{10} 2 = (x - 1) \log_{10} 3$
19. $\log_{10} (x - 3) = 1 + \log_{10} (x + 1)$
20. $\log_{10} (x + 2) = 1 - \log_{10} (x - 1)$

CHAPTER 12 REVIEW

1. Sketch the graphs of the following.
   (a) $y = 2^x$ (b) $y = 10^{-x}$
2. Write the following in logarithmic notation.
   (a) $8^{2/3} = 2$ (b) $(16)^{3/4} = 8$
3. Write in exponential notation.
   (a) $\log_3 81 = 4$ (b) $\log_{10} 0.001 = -3$
4. Find:
   (a) $\log_{10} 4.64$
   (b) $\log_{10} 0.0321$
   (c) $\log_{10} 0.000111$
   (d) $\log_{10} 345{,}000$
5. Find:
   (a) $\text{antilog}_{10}\ 1.5752$
   (b) $\text{antilog}_{10}\ (9.9689 - 10)$
   (c) $\text{antilog}_{10}\ 3.8069$
   (d) $\text{antilog}_{10}\ (8.1072 - 10)$

Express as the sum or difference of simpler logarithmic quantities (Exercises 6–11).

6. $\log_{10} x^2 y$
7. $\log_{10} \dfrac{x^3}{3y}$
8. $\log_{10} \dfrac{x\sqrt{y}}{z^3}$
9. $\log_{10} \sqrt[5]{x}$
10. $\log_{10} \dfrac{x^3 \sqrt{y}}{\sqrt{z}}$
11. $\log_{10} \dfrac{3x^2 y}{\sqrt[4]{z^3}}$

Compute, using logarithms (Exercises 12–15).

12. $(3.42)(11.7)$
13. $\dfrac{78.96}{12.1}$
14. $\sqrt[3]{14.7}$
15. $\dfrac{(9.61)(4.03)}{\sqrt{8.64}}$

EXPONENTIAL AND LOGARITHMIC FUNCTIONS

Find the solution sets (Exercises 16–17).

16. $3^x = 2$ 17. $7^{2x} = 4^{x-1}$

Find the solution sets of each of the following (Exercises 18–20).

18. $\log_x 32 = 5$ 19. $\log_6 x = 0$ 20. $\log_{\sqrt{2}} 8 = x$

APPENDIX:
Four Place Logarithms of Numbers

N	0	1	2	3	4	5	6	7	8	9
10	0000	0043	0086	0128	0170	0212	0253	0294	0334	0374
11	0414	0453	0492	0531	0569	0607	0645	0682	0719	0755
12	0792	0828	0864	0899	0934	0969	1004	1038	1072	1106
13	1139	1173	1206	1239	1271	1303	1335	1367	1399	1430
14	1461	1492	1523	1553	1584	1614	1644	1673	1703	1732
15	1761	1790	1818	1847	1875	1903	1931	1959	1987	2014
16	2041	2068	2095	2122	2148	2175	2201	2227	2253	2279
17	2304	2330	2355	2380	2405	2430	2455	2480	2504	2529
18	2553	2577	2601	2625	2648	2672	2695	2718	2742	2765
19	2788	2810	2833	2856	2878	2900	2923	2945	2967	2989
20	3010	3032	3054	3075	3096	3118	3139	3160	3181	3201
21	3222	3243	3263	3284	3304	3324	3345	3365	3385	3404
22	3424	3444	3464	3483	3502	3522	3541	3560	3579	3598
23	3617	3636	3655	3674	3692	3711	3729	3747	3766	3784
24	3802	3820	3838	3856	3874	3892	3909	3927	3945	3962
25	3979	3997	4014	4031	4048	4065	4082	4099	4116	4133
26	4150	4166	4183	4200	4216	4232	4249	4265	4281	4298
27	4314	4330	4346	4362	4378	4393	4409	4425	4440	4456
28	4472	4487	4502	4518	4533	4548	4564	4579	4594	4609
29	4624	4639	4654	4669	4683	4698	4713	4728	4742	4757
30	4771	4786	4800	4814	4829	4843	4857	4871	4886	4900
31	4914	4928	4942	4955	4969	4983	4997	5011	5024	5038
32	5051	5065	5079	5092	5105	5119	5132	5145	5159	5172
33	5185	5198	5211	5224	5237	5250	5263	5276	5289	5302
34	5315	5328	5340	5353	5366	5378	5391	5403	5416	5428
35	5441	5453	5465	5478	5490	5502	5514	5527	5539	5551
36	5563	5575	5587	5599	5611	5623	5635	5647	5658	5670
37	5682	5694	5705	5717	5729	5740	5752	5763	5775	5786
38	5798	5809	5821	5832	5843	5855	5866	5877	5888	5899
39	5911	5922	5933	5944	5955	5966	5977	5988	5999	6010
40	6021	6031	6042	6053	6064	6075	6085	6096	6107	6117
41	6128	6138	6149	6160	6170	6180	6191	6201	6212	6222
42	6232	6243	6253	6263	6274	6284	6294	6304	6314	6325
43	6335	6345	6355	6365	6375	6385	6395	6405	6415	6425
44	6435	6444	6454	6464	6474	6484	6493	6503	6513	6522
45	6532	6542	6551	6561	6571	6580	6590	6599	6609	6618
46	6628	6637	6646	6656	6665	6675	6684	6693	6702	6712
47	6721	6730	6739	6749	6758	6767	6776	6785	6794	6803
48	6812	6821	6830	6839	6848	6857	6866	6875	6884	6893
49	6902	6911	6920	6928	6937	6946	6955	6964	6972	6981
50	6990	6998	7007	7016	7024	7033	7042	7050	7059	7067
51	7076	7084	7093	7101	7110	7118	7126	7135	7143	7152
52	7160	7168	7177	7185	7193	7202	7210	7218	7226	7235
53	7243	7251	7259	7267	7275	7284	7292	7300	7308	7316
54	7324	7332	7340	7348	7356	7364	7372	7380	7388	7396

N	0	1	2	3	4	5	6	7	8	9
55	7404	7412	7419	7427	7435	7443	7451	7459	7466	7474
56	7482	7490	7497	7505	7513	7520	7528	7536	7543	7551
57	7559	7566	7574	7582	7589	7597	7604	7612	7619	7627
58	7634	7642	7649	7657	7664	7672	7679	7686	7694	7701
59	7709	7716	7723	7731	7738	7745	7752	7760	7767	7774
60	7782	7789	7796	7803	7810	7818	7825	7832	7839	7846
61	7853	7860	7868	7875	7882	7889	7896	7903	7910	7917
62	7924	7931	7938	7945	7952	7959	7966	7973	7980	7987
63	7993	8000	8007	8014	8021	8028	8035	8041	8048	8055
64	8062	8069	8075	8082	8089	8096	8102	8109	8116	8122
65	8129	8136	8142	8149	8156	8162	8169	8176	8182	8189
66	8195	8202	8209	8215	8222	8228	8235	8241	8248	8254
67	8261	8267	8274	8280	8287	8293	8299	8306	8312	8319
68	8325	8331	8338	8344	8351	8357	8363	8370	8376	8382
69	8388	8395	8401	8407	8414	8420	8426	8432	8439	8445
70	8451	8457	8463	8470	8476	8482	8488	8494	8500	8506
71	8513	8519	8525	8531	8537	8543	8549	8555	8561	8567
72	8573	8579	8585	8591	8597	8603	8609	8615	8621	8627
73	8633	8639	8645	8651	8657	8663	8669	8675	8681	8686
74	8692	8698	8704	8710	8716	8722	8727	8733	8739	8745
75	8751	8756	8762	8768	8774	8779	8785	8791	8797	8802
76	8808	8814	8820	8825	8831	8837	8842	8848	8854	8859
77	8865	8871	8876	8882	8887	8893	8899	8904	8910	8915
78	8921	8927	8932	8938	8943	8949	8954	8960	8965	8971
79	8976	8982	8987	8993	8998	9004	9009	9015	9020	9025
80	9031	9036	9042	9047	9053	9058	9063	9069	9074	9079
81	9085	9090	9096	9101	9106	9112	9117	9122	9128	9133
82	9138	9143	9149	9154	9159	9165	9170	9175	9180	9186
83	9191	9196	9201	9206	9212	9217	9222	9227	9232	9238
84	9243	9248	9253	9258	9263	9269	9274	9279	9284	9289
85	9294	9299	9304	9309	9315	9320	9325	9330	9335	9340
86	9345	9350	9355	9360	9365	9370	9375	9380	9385	9390
87	9395	9400	9405	9410	9415	9420	9425	9430	9435	9440
88	9445	9450	9455	9460	9465	9469	9474	9479	9484	9489
89	9494	9499	9504	9509	9513	9518	9523	9528	9533	9538
90	9542	9547	9552	9557	9562	9566	9571	9576	9581	9586
91	9590	9595	9600	9605	9609	9614	9619	9624	9628	9633
92	9638	9643	9647	9652	9657	9661	9666	9671	9675	9680
93	9685	9689	9694	9699	9703	9708	9713	9717	9722	9727
94	9731	9736	9741	9745	9750	9754	9759	9763	9768	9773
95	9777	9782	9786	9791	9795	9800	9805	9809	9814	9818
96	9823	9827	9832	9836	9841	9845	9850	9854	9859	9863
97	9868	9872	9877	9881	9886	9890	9894	9899	9903	9908
98	9912	9917	9921	9926	9930	9934	9939	9943	9948	9952
99	9956	9961	9965	9969	9974	9978	9983	9987	9991	9996

Answers to Odd-Numbered Problems

EXERCISES 1.1

1. {Sunday, Monday, Tuesday, Wednesday, Thursday, Friday, Saturday}.
3. {Winter, Spring, Summer, Fall}.
5. {1, 2, 3, 4, 5, 6, 7, 8, 9}.
7. {−4, −3, −2, −1, 0, 1, 2, 3, 4}.
9. {2, 4, 6, 8, ...}.
11. $\{x \mid x \text{ is a taxpayer}\}$.
13. $\{x \mid x \text{ is a Senator of the United States}\}$.
15. $\{x \mid x \text{ is an integer and } x \text{ is greater than } -3\}$.
17. $\{x \mid x \text{ is a natural number and } x \text{ is greater than } 12\}$.
19. $\{x \mid x \text{ is an integer and } x \neq 4\}$.
21. Finite.   23. Finite.   25. Infinite.   27. Finite.   29. Infinite.
31. {1, 2, 3}.   33. {January, June, July}.
35. {16, 17, 18, 19, 20, 21, 22, 23, 24, 25, 26, 27, 28, 29}.
37. Multiplication property.
39. Substitution property.
41. Subtraction property.
43. Multiplication cancellation property.

EXERCISES 1.2

1. Closure property for multiplication.
3. Closure property for addition.
5. Distributive property.
7. Multiplicative inverse.   9. Identity element for addition.
11. Identity element for addition.   13. Associative property of addition.
15. Commutative property of multiplication.
17. Commutative property of multiplication.   19. Multiplicative inverse.
21. Additive inverse.
23. True.   25. True.   27. True.   29. False.   31. True.   33. False.

35. (a) $\frac{1}{7}$; (b) -2; (c) $\frac{1}{x}$; (d) $\frac{1}{a+b}$; (e) $\frac{1}{x^2-t^2}$; (f) $\frac{y}{3x}$; (g) $\frac{1}{5x-2y}$;
(h) $\frac{1}{\sqrt{5}}$ or $\frac{\sqrt{5}}{5}$
37. $<$.   **39.** $<$.   **41.** $<$.   **43.** $>$.   **45.** $>$.   **47.** $=$.   **49.** $<$.
51. $=$.

EXERCISES 1.3

1. 2.   **3.** $(-1)(45)$.   **5.** $(-1)(34)$.   **7.** $(-1)(516)$.   **9.** $2^3 \cdot 3^2$.
11. 5^3.   **13.** 7^2.   **15.** $2^2 \cdot 3 \cdot 5$.   **17.** $2^3 \cdot 5^3$.   **19.** $13 \cdot 23$.

EXERCISES 1.4

1. 6.   **3.** -13.   **5.** -13.   **7.** -5.   **9.** 5.   **11.** 3.   **13.** 0.
15. -45.   **17.** -9.   **19.** -38.   **21.** -77.   **23.** -66.
25. (a) $6 + 7$;   (b) $8 + (-15)$;   (c) $-12 + 4$;   (d) $-16 + (-8)$;
(e) $-20 + 18$;   (f) $72 + (-36)$.
27. Negative.   **29.** Negative.   **31.** $23°$.
33. (a) 13,966 feet;   (b) 12,492 feet;   (c) 12,812 feet.
35. (a) 3;   (b) $-\frac{1}{2}$;   (c) $\frac{5}{6}$;   (d) $\frac{8}{5}$;   (e) $-\frac{3}{3}$ or -1;   (f) 10;   (g) -21;
(h) $\frac{3}{3}$ or 1;   (i) $-\frac{7}{21}$ or $-\frac{1}{3}$;   (j) $\frac{2}{5}$.
37. $=$.   **39.** $<$.   **41.** $=$.   **43.** $=$.   **45.** $<$.

EXERCISES 1.5

1. 14.   **3.** 45.   **5.** 54.   **7.** -24.   **9.** -3.   **11.** -32.   **13.** 60.
15. -21.   **17.** 12.   **19.** 0.   **21.** -30.   **23.** 3.   **25.** 4.
27. -25.   **29.** 0.   **31.** -18.   **33.** -38.   **35.** 7.
37. (a) -9;   (b) 5;   (c) 10;   (d) -9.
39. $\frac{2}{1}$.   **41.** $\frac{15}{8}$.   **43.** $\frac{18}{16}$ or $\frac{9}{8}$.   **45.** 8.   **47.** -5.   **49.** -6.
51. $\frac{2}{3}$.   **53.** $-\frac{24}{35}$.   **55.** $\frac{33}{14}$.   **57.** $=$.   **59.** $=$.

CHAPTER 1 REVIEW

1. (a) $\{7, 8, 9, \ldots\}$;   (b) $\{1, 3, 5, 7, \ldots\}$;
(c) $\{3, 4, 5, 6, 7, 8, 9, 10, 11\}$;   (d) $\{\ldots, -5, -4, -3\}$.
3. (a) Commutative property of addition;   (b) multiplicative inverse;
(c) distributive property;   (d) additive inverse.
5. 7.   **7.** 24.   **9.** -114.   **11.** 6.   **13.** $\frac{1}{\sqrt{3}}$ or $\frac{\sqrt{3}}{3}$ **15.** 1.
17. True.   **19.** False.

EXERCISES 2.1

1. 8.   3. -1.   5. $-\frac{1}{8}$.   7. 1000.   9. 0.01.   11. -16.
13. (a) Positive; (b) positive; (c) positive; (d) negative; (e) negative; (f) positive; (g) positive; (h) positive; (i) negative; (j) positive.
15. a^7.   17. z^7.   19. $\frac{1}{y^3}$.   21. x^9.   23. $x^4y^4z^6$.   25. $\frac{81x^{12}}{y^4}$.
27. $\frac{a^9}{b^3c^6}$.   29. $\frac{16y^8}{x^{12}}$.   31. $512a^{21}$.   33. $\frac{648a^{11}}{b^{15}}$.   35. $\frac{1}{2a^6x}$.
37. $-32y^5$.   39. $-\frac{576x^5z^{14}}{y^{17}}$.   41. x^{4nr}.   43. y^{9k}.   45. x^2.
47. x^{2b}.   49. x^n.   51. $\frac{(a+b)^4}{(a-b)^4}$.   53. y^{2n+4}.

EXERCISES 2.2

1. 8.   3. -3.   5. $|x|$.   7. xy.   9. 2.   11. $\frac{x^8}{4}$.   13. $|x^3|$.
15. -12.   17. $2x^2$.   19. $3|x|y^2$.   21. $\frac{-2}{5}y^2$.   23. $\frac{12}{17}y^2$.
25. $\frac{-2x^2}{y^2z}$.

EXERCISES 2.3

1. 2.   3. 2.   5. 0.2.   7. $\frac{4}{9}$.   9. 27.   11. 9.   13. $\sqrt{x}$.
15. $\sqrt[8]{x^3}$ or $(\sqrt[8]{x})^3$.   17. $\sqrt[3]{5^2}$ or $(\sqrt[3]{5})^2$.   19. $\sqrt[3]{ax}$.
21. $\sqrt[4]{(ab)^3}$ or $\sqrt[4]{a^3b^3}$.   23. $\sqrt{(10x)^3}$ or $\sqrt{10^3 \cdot x^3}$.   25. $3\sqrt{x}$.
27. $\sqrt[3]{(8x)^2}$.   29. $\sqrt[4]{x^2y}$.   31. $\sqrt[3]{(x^3y^4)^2}$ or $\sqrt[3]{x^6y^8}$.   33. $4x^2\sqrt[5]{(x^3y^4)^3}$.
35. $\sqrt[3]{(x+y)^2}$.   37. $\sqrt[3]{(3x+y)^2}$.   39. $\sqrt[5]{(2x+y)^3}$.   41. $8^{1/2}$.
43. $5^{1/3}$.   45. $y^{2/3}$.   47. $(3x^2)^{1/3}$ or $3^{1/3}x^{2/3}$.   49. $z^{2/3}$.
51. $(-x^2)^{1/3}$ or $-(x^{2/3})$.   53. $(-37x^4)^{1/3}$ or $(-37)^{1/3}x^{4/3}$ or $-(37)^{1/3}x^{4/3}$.
55. $(z^3y^5)^{1/4}$ or $z^{3/4}y^{5/4}$.   57. $-2(10x)^{1/2}$ or $-2(10)^{1/2}x^{1/2}$.
59. $(xy)^{1/3}$ or $x^{1/3}y^{1/3}$.   61. xy.   63. x.   65. $2^{7/4}$.   67. x^3.   69. y.
71. $-2xy^{2/3}$.   73. $\frac{1}{y^{1/3}}$.   75. $\frac{a^{5/3}}{b}$.   77. $\frac{1}{x^2y^{4/3}}$.   79. $-p^2$.
81. $\frac{256y^{8/3}}{x^{4/3}}$.

EXERCISES 2.4

1. $\dfrac{1}{x^2}$. 3. $\dfrac{1}{z^{1/3}}$. 5. $\dfrac{b}{a^4}$. 7. $\tfrac{3}{2}$. 9. $\dfrac{2c^3}{d^8}$. 11. $\dfrac{3x^2z^2}{y^3t^3}$.

13. $\dfrac{(3b)^2}{8a}$. 15. $\dfrac{1}{x}+\dfrac{1}{y}$ or $\dfrac{x+y}{xy}$. 17. $\dfrac{x+\dfrac{1}{x}}{y+\dfrac{1}{y}}$ or $\dfrac{y(x^2+1)}{x(y^2+1)}$.

19. $\dfrac{\dfrac{1}{a}-\dfrac{1}{b}}{2\left(\dfrac{1}{a}-\dfrac{1}{b}\right)}$ or $\dfrac{1}{2}$. 21. 4. 23. $6xy^3$. 25. $\dfrac{x}{y}$. 27. $x^{1/4}$.

29. $-6x^3$. 31. $\tfrac{1}{72}$. 33. $\dfrac{y^{10}}{9x^8}$. 35. $\dfrac{9x^4b^2}{4}$. 37. $\dfrac{1}{x}$. 39. $\dfrac{1}{x^{3n}y^{9n}}$.

EXERCISES 2.5

1. 4. 3. $2\sqrt{y}$. 5. $|y|\sqrt{x}$. 7. $2|x|\sqrt{10}$. 9. y. 11. $2x^2\sqrt{5}$.

13. $-3y\sqrt[3]{xy}$. 15. $10y\sqrt{xy}$. 17. $\dfrac{\sqrt{3}}{3}$. 19. $\sqrt{x}$. 21. $\sqrt[3]{2}$.

23. $\dfrac{\sqrt{2x}}{|x|}$. 25. $\dfrac{\sqrt{aby}}{|y|}$. 27. $\sqrt{2}$. 29. $\sqrt{|y|}$. 31. $\sqrt{3a}$.

33. $\sqrt{7|a|}$. 35. $\sqrt{yz^2}$ or $|z|\sqrt{y}$. 37. $3x\sqrt[3]{4x^2}$. 39. $\sqrt[3]{7x^2}$.

41. $\dfrac{2|y|}{|z|}\sqrt{2xz}$. 43. $\dfrac{x}{3}\sqrt[4]{3x^3y^2}$. 45. $\dfrac{5x^3y}{2}\sqrt[3]{2x}$. 47. $5-\sqrt{3}$.

49. $\dfrac{8(5\sqrt{3}+7)}{26}$ or $\dfrac{20\sqrt{3}+28}{13}$.

EXERCISE 2.6

1. $2\sqrt{2}$. 3. $9\sqrt{3}$. 5. $3\sqrt[3]{2}$. 7. $11\sqrt{3a}$. 9. $10x^2\sqrt{2x}$.

11. $\dfrac{9\sqrt{2}}{8}$. 13. $\dfrac{\sqrt{5}-8}{10}$. 15. 48. 17. 18. 19. 3. 21. 2.

23. $6x^4y^4\sqrt{xyz}$. 25. $15a^2b^4\sqrt{abc}$. 27. $2ab^2\sqrt[4]{2a^2b^3c^3}$.

29. $\sqrt{6}-2\sqrt{3}+4$. 31. $24+12\sqrt{3}$. 33. $8-2\sqrt{15}$. 35. $\sqrt{6}$.

37. $61-24\sqrt{5}$. 39. $2a-2\sqrt{a^2-b^2}$. 41. $\sqrt{2}$. 43. $\dfrac{|y|}{2x^2}\sqrt{2x}$.

45. $\dfrac{1}{y}\sqrt[5]{8y^4z^3}$. 47. $-\dfrac{2a}{b^2}\sqrt[5]{a^4b^2}$. 49. $\dfrac{|d|}{7c^2}\sqrt{14c}$.

EXERCISES 2.7

1. $2.68(10)^5$. **3.** $2.6(10)^{-7}$. **5.** $1.4(10)^7$. **7.** 32,000. **9.** 0.000316.
11. 0.0000000000000000000936. **13.** $1.97(10)^8$. **15.** $4.3(10)^9$.
17. 10^4. **19.** 10^{-3}. **21.** $1.944(10)^{11}$. **23.** $1.63(10)^6$.

CHAPTER 2 REVIEW

1. $\dfrac{x^6 y^8}{z^2}$. **3.** $\dfrac{-2x^2 y}{z^3}$. **5.** $\dfrac{x^{1/2} y^{1/3}}{z^{3/4}}$. **7.** $\dfrac{x}{y^6 z^{1/2}}$.

9. $\dfrac{\left(\dfrac{1}{a} - \dfrac{1}{b}\right)\left(\dfrac{1}{c} + \dfrac{1}{d}\right)}{\left(\dfrac{1}{a} + \dfrac{1}{b}\right)\left(\dfrac{1}{c} - \dfrac{1}{d}\right)}$ or $\dfrac{(b-a)(d+c)}{(b+a)(d-c)}$. **11.** $\dfrac{x\sqrt[3]{2xyz}}{z^2 y}$. **13.** $6xy\sqrt[3]{z^2}$.

15. $a + 9b - 6\sqrt{ab}$. **17.** $x^{1/2} - y^{1/2}$. **19.** $4.2(10)^4$.

EXERCISES 3.1

1. (a) Trinomial; degree 3; (b) binomial; degree 2; (c) monomial; degree 0;
(d) trinomial; degree 4; (e) binomial; degree 1; (f) trinomial; degree 2.
3. (a) 2; (b) 2; (c) 2; (d) 3; (e) 2; (f) 3. **5.** 1. **7.** -4. **9.** 4.
11. (a) 4; (b) 12; (c) 9; (d) $\frac{15}{4}$ or $3\frac{3}{4}$; (e) $\frac{75}{16}$ or $4\frac{11}{16}$; (f) 37.
13. True. **15.** True. **17.** False. **19.** False.

EXERCISES 3.2

1. $5x + 1$. **3.** $2x^3 - 2x$. **5.** $2x^2 y^2 + xy - 3$. **7.** $2a^3 + 3a^2 b + 2ab^2$.
9. $-2b^3$. **11.** $4x$. **13.** $x^2 + 1$. **15.** $2b^3$. **17.** $-2x^2 - 4x + 14$.
19. $-y^3 + 2$. **21.** $2 - 2bc$. **23.** $x^2 - 4x + 7$. **25.** $3x^2 y - 3xy^2$.
27. $t^2 - t$. **29.** $-2p^2 + 5p - 1$.

EXERCISES 3.3

1. $6xy$. **3.** $9x^3 y$. **5.** $-\frac{1}{6}r^3 s^3 t$. **7.** $6y^3 - 6y$. **9.** $-2x^2 y^2 - 2y^4$.
11. $6x^2 + 11x - 7$. **13.** $4x^2 + 4xy + y^2$. **15.** $a^2 - b^2$. **17.** $x^6 - y^6$.
19. $a^4 + 2a^3 b - 2ab^3 - b^4$.

EXERCISES 3.4

1. $a^2 - b^2$. **3.** $z^2 - 4z + 4$. **5.** $y^2 - 16$. **7.** $x^2 + 7x + 12$.
9. $t^2 - 7t + 12$. **11.** $a^4 - 6a^2 b^2 + 9b^4$. **13.** $25r^2 - 30r + 9$.
15. $9x^2 + 24xy + 16y^2$. **17.** $8x^2 - 26x + 15$. **19.** $x^2 + 15x + 50$.
21. $7x^2 - 26xy - 8y^2$. **23.** $81x^2 - 16y^2$. **25.** $64a^2 - 112ab + 49b^2$.

27. $10x^2 + 11ax + 3a^2$. **29.** $r^4 - s^4$. **31.** $16c^2 - 24cd + 9d^2$.
33. $49x^4y^4a^2 - 64$. **35.** $9x^2 + 6x - 35$. **37.** $24x^4 + 44x^2y^2 - 28y^4$.
39. 1369.

EXERCISES 3.5

1. $4x^2$. **3.** $-xy$. **5.** $-16t^2$. **7.** 0. **9.** x^ny^n. **11.** $3x + 6$.
13. $-2xy^2 + 3y$. **15.** $x + 2$. **17.** $5x^2y^2 - 3xy$. **19.** $a^3b^3 + 2a^2 + 2b$.
21. $-4xy^2 + 5y - 8x^3y^3$. **23.** $-9a^2b^2 + 8ab - 11$. **25.** $a^n + 2$.
27. $m - 2n + 5$.

EXERCISES 3.6

1. $y(y^2 + 3)$. **3.** $2y^3(1 - 3x)$. **5.** $5c(ab + 3bc - 2ac)$.
7. $-4ab^2(4b - a)$. **9.** $a(b - a + 2x)$. **11.** $3a(1 - 2ab + 3a^2)$.
13. $3f(1 - 4f^2 + 6f^3)$. **15.** $4(k^2 + 5k - 6)$ or $4(k + 6)(k - 1)$
17. $2x^2(2x - 4x^2y - 3y^2)$. **19.** $r^2(x^2 + 3r)$. **21.** $-4(3p^3 - 4p^2t + 12t^3)$.
23. $a(b^2 - 4ab^2 - a^2)$. **25.** $-16x^3(x^2 + y^4 + x)$.
27. $25ab(5a^2b^2 - ab - 8)$. **29.** $4ab(2 - ab - 12b^2)$. **31.** $(a + b)(x + y)$.
33. $(x - 1)(y + 3)$. **35.** $(a - x)(b - 3)$. **37.** $(y + z)(x - 3t)$.
39. $(x + 1)(5 + 3y)$. **41.** $(y + 1)(y^2 - 5)$. **43.** $(3p - q^2)(2m + 3n)$.
45. $(a^2 - 5)(a + 3)$. **47.** $(4a - 5d)(3b + c)$. **49.** $(x - y)(x - y - 1)$.

EXERCISES 3.7

1. $(x + 3)^2$. **3.** $(y - 5)^2$. **5.** $(x - 1)^2$. **7.** $(2x - 3y)^2$.
9. $(3x + 4)(3x - 4)$. **11.** $(10 + x)(10 - x)$. **13.** $(x + 5)(x - 5)$.
15. $(5 + 3y)(5 - 3y)$. **17.** $(x - 3)(x - 2)$. **19.** $(x + 7)(x - 4)$.
21. $(x - 4)(x - 7)$. **23.** $(y - 4)(y + 3)$. **25.** $(x - 4)(x + 1)$.
27. $(x + 5)(x - 4)$. **29.** $(3x + 1)(x + 3)$. **31.** $(2r - 5)(r - 1)$.
33. $(3x + 2)(x + 5)$. **35.** $(6x - 5)(x + 1)$. **37.** $(3x - 2)(4x - 3)$.
39. $(3y + 4)(y - 7)$. **41.** $(3x + 5)(x + 1)$. **43.** $(2x - 1)(x + 2)$.
45. $(8y^3 - 9z)(2y^3 - z)$. **47.** $(a^n + b^n)(a^n - b^n)$.
49. $(3x + 3y + 1)(x + y + 2)$.

EXERCISES 3.8

1. $(x - y)(x^2 + xy + y^2)$. **3.** $(y - 2)(y^2 + 2y + 4)$.
5. $(x - y^2)(x^2 + xy^2 + y^4)(x^6 + x^3y^6 + y^{12})$.
7. $(a^2 + b)(a^2 - b)(a^4 + a^2b + b^2)(a^4 - a^2b + b^2)$.
9. $8(xy^3 + 2z)(x^2y^6 - 2xy^3z + 4z^2)$. **11.** $(a^4 + 6b^2)(a^8 - 6a^4b^2 + 36b^4)$.
13. $(7 + 2x)(49 - 14x + 4x^2)$.
15. $(x + y + 3z)([x + y]^2 - 3z[x + y] + 9z^2)$.
17. $(5a + b - c - d)([5a + b]^2 + [5a + b][c + d] + [c + d]^2)$.
19. $[(n + 1) - (n - 2)][(n + 1)^2 + (n + 1)(n - 2) + (n - 2)^2]$ or $9(n^2 - n + 1)$.

ANSWERS TO ODD-NUMBERED PROBLEMS

CHAPTER 3 REVIEW

1. $2x^6 - 2x^5 - 2x^4 + x^3 + x^2 - 2x - 7$.   3. $8x^3 - 22x^2 + 19x - 5$.
5. $(3x + 2y)(3x - 2y)$.   7. $(2x + 5y)^2$.   9. $2(x - 3)(y + 2)$.
11. $(x + 1)(x + 2)$.   13. $(3x + 2)(x - 1)$.   15. $(2x + 3)(x + 1)$.
17. $(3x - 2)(2x + 5)$.   19. $(2x - 1)(4x^2 + 2x + 1)$.

EXERCISES 4.1

1. 0.   3. 1.   5. 2, 3.   7. -2.   9. x.   11. $\dfrac{1}{4x}$.   13. $\tfrac{1}{3}$.

15. $\dfrac{5}{x+1}$.   17. $x + 3$.   19. $\dfrac{x+2}{y-3}$.   21. $\dfrac{x^2 - 2x + 4}{x - 4}$.

23. $\dfrac{2r - 3}{2r - 5}$.   25. $\dfrac{1}{x+1}$.   27. $-\dfrac{1}{a+b}$.   29. $x - 3$.

EXERCISES 4.2

1. 24.   3. 42.   5. $6a^3b^2$.   7. $3x^3y^3$.   9. $a(x^2 - y^2)$.
11. $(x^2 - 4)(x + 3)$.   13. $(x - 1)(x - 2)(x - 4)$ 15. $2(x - 2)^2$.
17. $(x - 4)(x - 1)^3$.   19. $(x + y)(x^3 - y^3)$.

EXERCISES 4.3

1. True.   3. False.   5. True.   7. True.   9. True.   11. $\dfrac{10}{y}$.

13. $\dfrac{3a + 4b}{a - b}$.   15. $\dfrac{7a - 1}{a^2 + 2a + 1}$.   17. $\dfrac{-y - 2}{x^2 - x + 2}$.   19. $\dfrac{1 + x + y}{xy}$.

21. $\dfrac{b - x}{x^2 - ax}$.   23. $\dfrac{1}{xy}$.   25. $\dfrac{a^2 - 7}{a^2 - 9}$.   27. $2\left(\dfrac{x^2 + a^2}{x^2 - a^2}\right)$.

29. $\dfrac{x^2 - 2bx + x - a + b^2}{x^2 - b^2}$.   31. $\dfrac{x^2 - 3}{y}$.   33. $\dfrac{3x - 4y^2}{y^2}$.

35. $\dfrac{-x^2 + 3x + 2}{x^2 - 1}$.   37. $\dfrac{x^2 - 2x + 2}{(x - 2)^2}$.   39. $\dfrac{x + a}{ax}$.

EXERCISES 4.4

1. $\dfrac{x^3}{28}$.   3. $\dfrac{b^2}{a}$.   5. $xy(x - y)$.   7. $12y^2(y + 2)$.

9. $(x + 2)^2$.   11. $\dfrac{(x - 4)(x^2 + x + 1)}{x + 2}$.   13. $4(x^2 + 2x + 4)$.

15. $\dfrac{10(x + 2)}{(x - 4)(x + 5)}$.   17. $x + 1$.   19. $(x + y)^2$.

EXERCISES 4.5

1. $\dfrac{1}{y}$. 3. $\dfrac{9d}{2abc}$. 5. $\dfrac{15}{2}abc^3y$. 7. $\tfrac{1}{3}a^2b$. 9. $\dfrac{4(x+3)}{3(x+2)}$.

11. $\dfrac{6(5+y)}{5(2+y)}$. 13. $\dfrac{(y-4)(y+3)}{(y-1)(2y+1)}$. 15. $\dfrac{x+4}{x-4}$. 17. $2y-5$.

19. $\dfrac{(r+2)(r+1)}{(2r-1)(r-3)}$.

EXERCISES 4.6

1. $\dfrac{1}{2}c$ or $\dfrac{c}{2}$. 3. $\dfrac{y^2}{x}$. 5. $\dfrac{3(a+b)}{b^2}$. 7. $\dfrac{a-b}{a+b}$. 9. $\dfrac{x+2y}{x-y}$.

11. $\dfrac{x+y}{y-x}$. 13. $\dfrac{x+2}{x-3}$. 15. $-\dfrac{x^2+y^2}{4xy}$. 17. $-\dfrac{x}{y}$.

CHAPTER 4 REVIEW

1. (a) 0; (b) 2; (c) 3; (d) $-1, 1$. 3. $3(x+1)$. 5. $\dfrac{x+2}{x-7}$.

7. $\dfrac{2a-1}{2b}$. 9. $2\left[\dfrac{x^2+2x+2}{(x-1)(x+4)(x-4)}\right]$.

11. $\dfrac{-6y-x^2-2x-6}{(x+2)(x-2y)}$ or $-\dfrac{6y+x^2+2x+6}{(x+2)(x-2y)}$.

13. $\dfrac{2(3x+14y)}{(x+2y)(x-2y)(x+3y)}$. 15. $\dfrac{3a(2a+3)}{(a-1)(2a-1)}$. 17. $\dfrac{a-5}{a+5}$.

19. $3a$.

EXERCISES 5.1

1. Not a solution. 3. Solution. 5. Not a solution. 7. Solution.
9. $\{3\}$. 11. $\{4\}$. 13. $\{2\}$. 15. $\{7\}$. 17. $\{6\}$. 19. $\{\tfrac{1}{4}\}$.
21. $\{2\}$. 23. $\{\tfrac{7}{5}\}$. 25. $\{\tfrac{1}{3}\}$. 27. $\{8\}$. 29. $\{-\tfrac{11}{3}\}$. 31. $\{2\}$.
33. $\{7, -7\}$. 35. $\{4, -4\}$. 37. $\{1, -6\}$. 39. $\{24, -12\}$.
41. $\{-6\}$. 43. $\{-6\}$. 45. $\{-\tfrac{3}{8}\}$. 47. $\{1\}$.

EXERCISES 5.2

1. $x = 1 - b$. 3. $x = \dfrac{k}{a+b}$. 5. $x = \dfrac{a}{b}$. 7. $m = \dfrac{f}{a}$.

9. $g = \dfrac{2v}{t^2}$. 11. $W_1 = \dfrac{d_2 W_2}{d_1}$. 13. $W = \dfrac{8ds}{l^2}$. 15. $R = \dfrac{E}{I}$.

17. $C = \dfrac{5(F-32)}{9}.$ **19.** $g = \dfrac{2(at-S)}{t^2}.$ **21.** $a = \dfrac{S(1-r)}{1+r^n}.$

23. $n = \dfrac{1-a+d}{d}.$ **25.** $r_1 = \dfrac{Rr_2}{r_2-R}.$

EXERCISES 5.4

1. $\{x \mid x > -3\}.$

3. $\{x \mid x > -9\}.$

5. $\{x \mid x \le 2\}.$

7. $\{x \mid x > 4\}.$

9. $\{x \mid x > -2\}.$

11. $\{x \mid 6 \le x \le 12\}.$

13. $\{x \mid x \le 7\}.$

15. $\{x \mid x < -4\}.$

17. $\{x \mid x \le 1\}.$

19. $\{x \mid x > 5 \text{ or } x < -3\}.$

21. $\{x \mid -4 \le x \le 3\}.$

23. $\{x \mid x \ge 7 \text{ or } x \le 1\}.$

25. $\{x \mid -7 \le x \le 8\}.$

EXERCISES 5.5

1. 39, 41, 43. 3. 18, 42. 5. 12. 7. 12, 13. 9. 16.
11. 16 quarters, 32 dimes, 80 nickels. 13. $432 at $6\frac{1}{2}$%, $468 at 6%.
15. 300. 17. 40 pounds clover, 60 pounds grass. 19. 110 gallons.
21. 6 days. 23. 18 minutes. 25. $6\frac{3}{7}$ days. 27. 11:30 a.m.
29. 50 mph, 58 mph. 31. $12,000. 33. 95. 35. 42 37. 999.

CHAPTER 5 REVIEW

1. {3}. 3. {17}. 5. $h = \dfrac{3V}{4\pi r^2}$.

7. $\{x \mid x < 1\}$.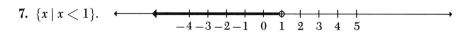

9. $\{x \mid -3 \leq x \leq -2\}$.

11. $\{x \mid x < 2 \text{ or } x > 4\}$.
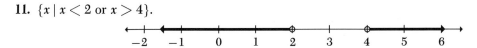

13. $\{x \mid x < -1 \text{ or } x > 5\}$.

15. $\{x \mid x \leq 1 \text{ or } x \geq 3\}$.

17. 20, 22, 24. 19. 2 quarts.

EXERCISES 6.1

1. $\{(-1, -1), (-1, 0), (-1, 1), (0, -1), (0, 0), (0, 1), (1, -1), (1, 0), (1, 1)\}$.

3.

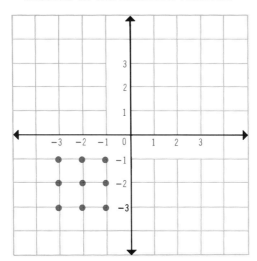

5. (a) 1; (b) 4; (c) 16; (d) n^2.   **7.** True.   **9.** False.

EXERCISES 6.2

1. First.   **3.** Fourth.   **5.** Third.   **7.** Third.   **9.** Fourth.   **11.** 0.
13. (a) $x > 0$ and $y > 0$; (b) $x < 0$ and $y > 0$; (c) $x < 0$ and $y < 0$;
(d) $x > 0$ and $y < 0$.   **15.** First or second quadrant.
17. Third or fourth quadrant.

19.

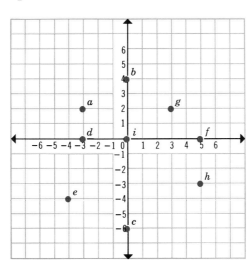

EXERCISES 6.3

1. -4.   **3.** -5.   **5.** 6.   **7.** 1.   **9.** 7.

11.

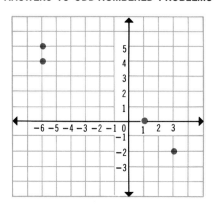

13.

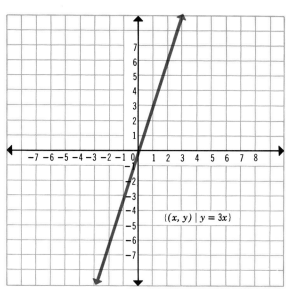

15.

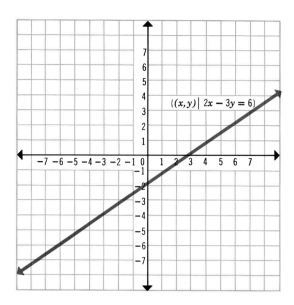

17.

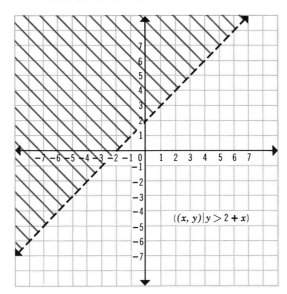

19.

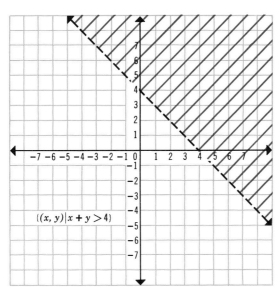

21. (a) A function; (b) not a function; (c) not a function; (d) a function; (e) a function; (f) a function; (g) not a function; (h) a function.
23. $D = \{1, 2, 3, 4\}; R = \{1, 2, 3, 4\}$.
25. $D = \{\ldots, -2, -1, 0, 1, 2, \ldots\}; R = \{0\}$.   **27.** False.   **29.** True.
31. False.   **33.** True.   **35.** True.

EXERCISES 6.4

1. (a) True; (b) True; (c) False; (d) False; (e) False; (f) True; (g) False; (h) True.   **3.** $A: \{-3, -2, 2, 3\}; B: \{-3, -2, -1, 0, 1, 2\}; h: R; f: R$.

278   ANSWERS TO ODD-NUMBERED PROBLEMS

5.

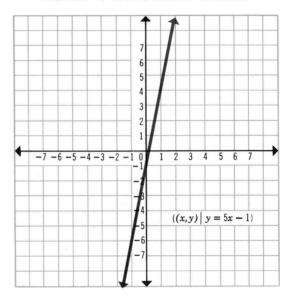

EXERCISES 6.5

1. a, c, e, f.

3.

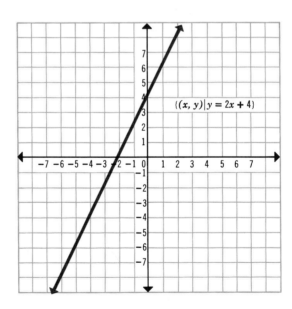

5.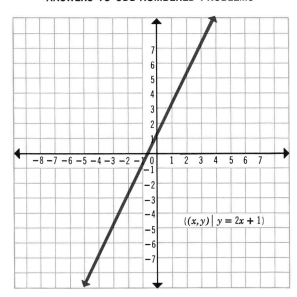

7.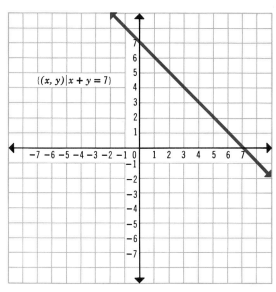

9.

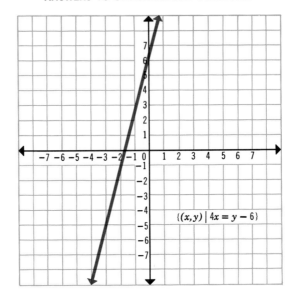

11. (a) 3; (b) -2; (c) $\frac{1}{2}$; (d) -1; (e) $\frac{3}{2}$; (f) -2; (g) $\frac{7}{6}$; (h) -2.

EXERCISES 6.6

1. $d = kt$.   **3.** $A = kh$.   **5.** $K = kV^2$.   **7.** $P = \dfrac{k}{V}$.   **9.** $P = kS$.

11. $B = \dfrac{k}{d^2}$.   **13.** $H = kd^3N$.   **15.** $F = k\dfrac{m_1 m_2}{d^2}$.   **17.** 392.   **19.** 5.

21. 3 cubic feet.   **23.** 45 miles.

CHAPTER 6 REVIEW

1. $\{(-2, -2), (-2, -3), (-2, -4), (-3, -2), (-3, -3), (-3, -4), (-4, -2), (-4, -3), (-4, -4)\}$.
3. (a) First; (b) fourth; (c) second; (d) fourth; (e) second; (f) fourth.
5. $D = \{-4, -3, -2, 4\}$; $R = \{3, 6, 7, 9\}$.   **7.** 0.   **9.** $\frac{1}{6}$.   **11.** 26.
13. x-intercept is 3; y-intercept is -9.   **15.** $\frac{3}{2}$.   **17.** $T = ks$.

19. x-intercept is 6; y-intercept is 2; slope is $-\frac{1}{3}$.

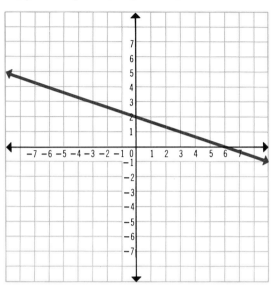

EXERCISES 7.1

1. $a = -3, b = 2$. **3.** $a = 3, b = -10$. **5.** $a = -15, b = 3$.
7. $a = 5, b = 3$. **9.** $a = -2, b = -\frac{3}{2}$. **11.** $(11, -6)$.
13. $(\frac{3}{4}, -\frac{3}{4})$. **15.** $(3\sqrt{2}, -3)$. **17.** $(1, 1)$. **19.** $(6, 5)$. **21.** $(1, 1)$.
23. $(3, -3)$. **25.** $(-8, 0)$. **27.** $(-3, -4)$. **29.** $(-6, 4)$.
31. $(-2, -5)$. **33.** $(-8, 4)$. **35.** $(0, 0)$. **37.** $(-x - y, -x + y)$.
39. $(-\sqrt{3}, 7)$.

EXERCISES 7.2

1. $(2, 3)$. **3.** $(13, 0)$. **5.** $(7, -2)$. **7.** $(0, 53)$. **9.** $(-3, 4)$.
11. $(25, 0)$. **13.** $(-50, -38)$. **15.** $(62, 10)$. **17.** $(0, 21)$.
19. $(-24, 0)$.

EXERCISES 7.3

1. $(\frac{1}{5}, -\frac{2}{5})$. **3.** $(\frac{3}{13}, \frac{2}{13})$. **5.** $(\frac{1}{5}, 0)$. **7.** $(\frac{1}{3}, 0)$. **9.** $(\frac{3}{26}, -\frac{1}{13})$.
11. $(6, 17)$. **13.** $(12, 9)$. **15.** $(2, -26)$. **17.** $(12, 18)$. **19.** $(0, 0)$.
21. $(2, -1)$. **23.** $(-\frac{4}{13}, \frac{6}{13})$. **25.** $(\frac{3}{4}, 1)$. **27.** $(0, 6)$. **29.** $(\frac{6}{5}, \frac{12}{5})$.

EXERCISES 7.4

1. (a) $6 + 3i$; (b) $7i$; (c) 4; (d) $-3 - 7i$; (e) $-\sqrt{2}i$; (f) $-\sqrt{3}$;
(g) $\sqrt{2} + 2i$; (h) $\frac{1}{2} - \frac{1}{4}i$; (i) $3\sqrt{2} - 7i$.

ANSWERS TO ODD-NUMBERED PROBLEMS

3. (a) $3 - 2i$; (b) $2 + 4i$; (c) $1 - i$; (d) $\sqrt{2} + 2i$; (e) $\frac{1}{2} + \frac{1}{4}i$;
(f) $3\sqrt{2} - \sqrt{6}i$; (g) $5 + 7i$; (h) $6 + 2\sqrt{2}i$.   5. $9 + 3i$.
7. $6 + 8i$.   9. $-4 + 7i$.   11. $-3 - 9i$.   13. $3 + 4i$.
15. $-1 + 3i$.   17. $4 + 8i$.   19. $16 - 30i$.   21. 7.   23. $26 + 2\sqrt{2}i$.
25. $3 + 4i$.   27. $-7 + 6\sqrt{2}i$.   29. $-28 + 96i$.   31. $\frac{5}{2} + \frac{3}{4}i$.
33. $\frac{24}{25} - \frac{18}{25}i$.   35. $-\frac{1}{2} + \frac{5}{2}i$.   37. $-\frac{9}{41} - \frac{40}{41}i$.   39. $\frac{5}{13} + \frac{12}{13}i$.
41. (a) -1; (b) $-i$; (c) 1; (d) 1; (e) -1; (f) 1; (g) 1; (h) i;
(i) -1; (j) $-i$.

EXERCISES 7.5

1. $\sqrt{5}i$.   3. $-\sqrt{3}i$.   5. $2\sqrt{2}i$.   7. $3\sqrt{3}i$.   9. $-2\sqrt{6}i$.
11. $-\sqrt{35}i$.   13. $-4\sqrt{2}i$.   15. $\sqrt{21}i$.   17. 0.   19. $5\sqrt{2}i$.

CHAPTER 7 REVIEW

1. (a) $a = -7, b = 1$; (b) $a = -3, b = 3$; (c) $a = 0, b = 5$;
(d) $a = 3, b = -4$; (e) $a = 2, b = -5$; (f) $a = 5, b = -7$.
3. $-1 + 7i$.   5. $18 + i$.   7. $2i$.   9. $-1 - i$.
11. (a) Real part 3; imaginary part -4; (b) real part 2; imaginary part 5;
(c) real part 3; imaginary part 0; (d) real part 0; imaginary part $-\sqrt{5}$.
13. $-3 - 2i$.   15. $2 - 11i$.   17. $\frac{8}{13} - \frac{1}{13}i$.   19. $-\frac{7}{11} + \frac{6\sqrt{2}}{11}i$.

EXERCISES 8.1

1. Quadratic.   3. Quadratic.   5. Quadratic.   7. Quadratic.
9. Quadratic.

EXERCISES 8.2

1.

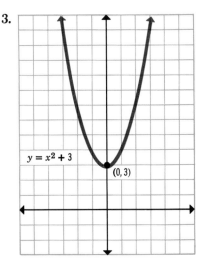

Minimum at (0, 0)

3.

Minimum at (0, 3)

5.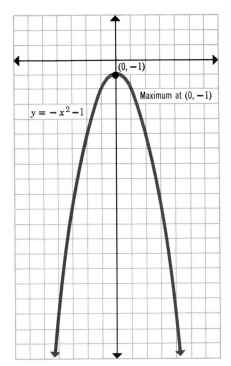

7.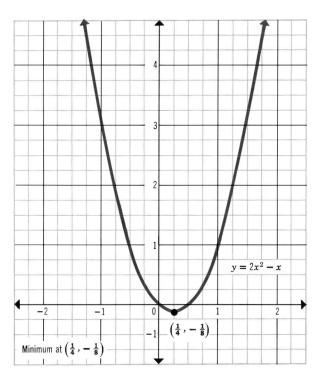

9.

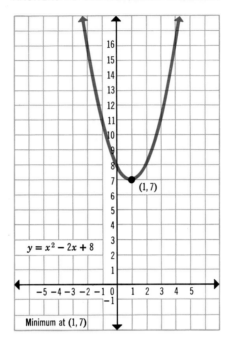

11. (a) Up; (b) up; (c) down; (d) down; (e) up; (f) down; (g) up; (h) down.

EXERCISES 8.3

1. 1.   **3.** 9.   **5.** 25.   **7.** $\frac{25}{4}$.   **9.** $\frac{49}{4}$.   **11.** 1.   **13.** 48.
15. -4.   **17.** -4.   **19.** $\frac{4}{3}$.   **21.** $y = (x + 6)^2$.   **23.** $y = (x + 3)^2 - 4$.
25. $y = -(x + 2)^2 + 4$.   **27.** $y = 2(x - 3)^2 + 1$.
29. $y = -2(x + 4)^2 - 1$.

31.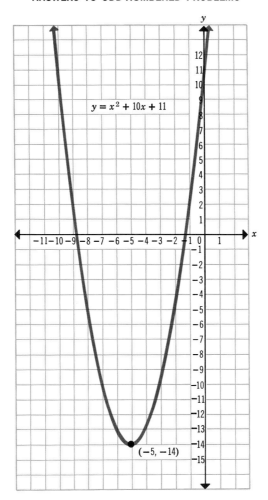

33.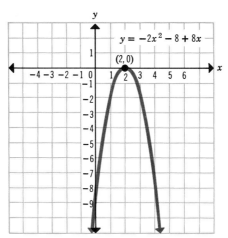

35.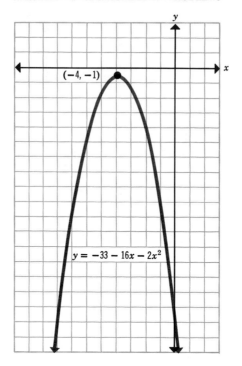

EXERCISES 8.4

1. 8, 8.   **3.** 8 feet by 8 feet.   **5.** 40 feet by 40 feet.
7. 625 square inches.   **9.** 275 watts.

CHAPTER 8   REVIEW

1. 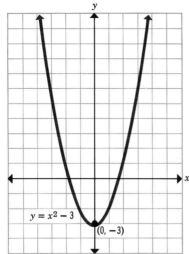   **3.**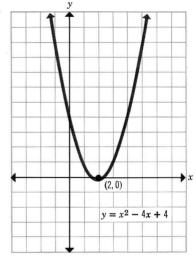

5.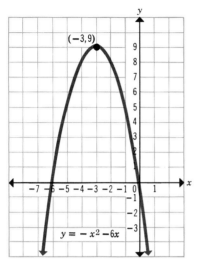

7.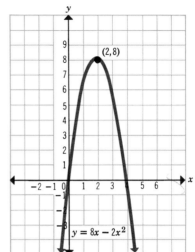

9.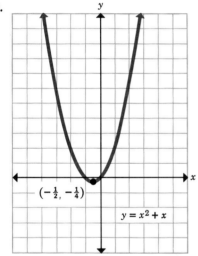

11.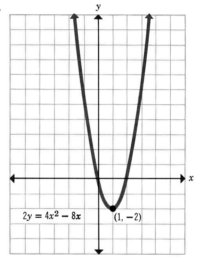

13.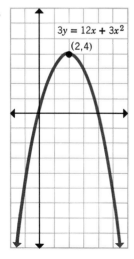

15.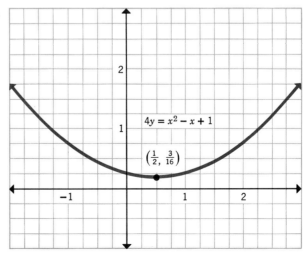

17. 400 square feet.　　**19.** In 5 seconds; 100 feet.

EXERCISES 9.1

1. $x^2 - 3x - 4 = 0$; $a = 1, b = -3, c = -4$.
3. $2x^2 - 6 = 0$; $a = 2, b = 0, c = -6$.
5. $x^2 - 3x - 9 = 0$; $a = 1, b = -3, c = -9$.
7. $x^2 + 5x - 7 = 0$; $a = 1, b = 5, c = -7$.
9. $x^2 - x - 5 = 0$; $a = 1, b = -1, c = -5$.　　**11.** $\{5, -5\}$.
13. $\left\{\sqrt{\frac{1}{2}}, -\sqrt{\frac{1}{2}}\right\}$ or $\left\{\frac{\sqrt{2}}{2}, -\frac{\sqrt{2}}{2}\right\}$.　　**15.** $\{1, -1\}$.　　**17.** $\{6, -6\}$.
19. $\{0, 3\}$.　　**21.** $\{0, -7\}$.　　**23.** $\{\sqrt{12}, -\sqrt{12}\}$ or $\{2\sqrt{3}, -2\sqrt{3}\}$.
25. $\{3, -3\}$.　　**27.** $\{0, \frac{1}{3}\}$.　　**29.** $\left\{\sqrt{\frac{7}{2}}i, -\sqrt{\frac{7}{2}}i\right\}$ or $\left\{\frac{\sqrt{14}}{2}i, -\frac{\sqrt{14}}{2}i\right\}$.
31. $\{-2, 5\}$.　　**33.** $\{\frac{1}{2}\}$.　　**35.** $\{2, \frac{1}{2}\}$.　　**37.** $\{5, -4\}$.　　**39.** $\{1, -8\}$.
41. $\{-\frac{1}{2}, 3\}$.　　**43.** $\{2, -1\}$.　　**45.** $\{-6, 2\}$.　　**47.** $\{\frac{1}{2}, -\frac{5}{3}\}$.　　**49.** $\{1, \frac{5}{8}\}$.

EXERCISES 9.2

1. $\{1 - 2\sqrt{2}, 1 + 2\sqrt{2}\}$.　　**3.** $\{-2 + 4\sqrt{2}, -2 - 4\sqrt{2}\}$.
5. $\{\frac{3}{2} + \frac{3}{2}\sqrt{3}, \frac{3}{2} - \frac{3}{2}\sqrt{3}\}$.　　**7.** $\left\{\frac{5}{3} + \frac{\sqrt{2}}{3}, \frac{5}{3} - \frac{\sqrt{2}}{3}\right\}$.
9. $\left\{\frac{1}{2} + \frac{\sqrt{3}}{2}i, \frac{1}{2} - \frac{\sqrt{3}}{2}i\right\}$.　　**11.** $\{2i, -2i\}$.
13. $\left\{-\frac{1}{2} + \frac{\sqrt{3}}{6}i, -\frac{1}{2} - \frac{\sqrt{3}}{6}i\right\}$.　　**15.** $\left\{-\frac{3}{2} + \frac{3\sqrt{5}}{2}, -\frac{3}{2} - \frac{3\sqrt{5}}{2}\right\}$.
17. $\{1, -\frac{9}{8}\}$.　　**19.** $\{\frac{1}{2} + \frac{1}{2}i, \frac{1}{2} - \frac{1}{2}i\}$.　　**21.** $\{1, -\frac{5}{9}\}$.　　**23.** $\{\frac{1}{8}\}$.
25. $\{-1, \frac{2}{3}\}$.　　**27.** $\{3, -\frac{3}{2}\}$.　　**29.** $\left\{\frac{k}{6} + \frac{|k|\sqrt{47}i}{6}, \frac{k}{6} - \frac{|k|\sqrt{47}i}{6}\right\}$.
31. Two complex.　　**33.** Two complex.　　**35.** Two distinct real.
37. Two distinct real.　　**39.** One real.　　**41.** $4, -4$.　　**43.** $2\sqrt{6}, -2\sqrt{6}$.
45. $\frac{1}{12}$.

EXERCISES 9.3

1. Sum 2, product 3.　　**3.** Sum 6, product 1.　　**5.** Sum -1, product 8.
7. Sum $-\frac{5}{3}$, product $\frac{1}{3}$.　　**9.** Sum $\frac{7}{5}$, product $-\frac{1}{5}$.　　**11.** $x^2 - 7x + 10 = 0$.
13. $x^2 - \frac{5}{6}x + \frac{1}{6} = 0$ or $6x^2 - 5x + 1 = 0$.
15. $x^2 + \frac{5}{2}x - \frac{3}{2} = 0$ or $2x^2 + 5x - 3 = 0$.
17. $12x^2 + x - 6 = 0$ or $x^2 + \frac{1}{12}x - \frac{1}{2} = 0$.　　**19.** $x^2 - 2 = 0$.

ANSWERS TO ODD-NUMBERED PROBLEMS

EXERCISES 9.4

1. 2. 3. $-4, 2$. 5. $0, -2$. 7. $\{5\}$. 9. $\{\sqrt{3}, -\sqrt{3}\}$.
11. $\{2\}$. 13. $\{2, \frac{11}{9}\}$. 15. $\{-7\}$. 17. $\{1, \frac{6}{5}\}$.

EXERCISES 9.5

1. $\{5\}$. 3. $\{0\}$. 5. $\{\frac{13}{4}\}$. 7. $\{\frac{1}{2}\}$. 9. $\{\frac{1}{8}\}$.

EXERCISES 9.6

1. $\{1, -1, \sqrt{2}, -\sqrt{2}\}$. 3. $\{1, -1, 3, -3\}$.
5. $\{\sqrt{3}, -\sqrt{3}, \sqrt{6}, -\sqrt{6}\}$. 7. $\{1, 256\}$. 9. $\{\frac{1}{4}, 16\}$. 11. $\{\frac{1}{4}, 16\}$.
13. $\{1, -1, \frac{1}{2}, -\frac{1}{2}\}$. 15. $\{\frac{1}{2}, -\frac{1}{2}\}$. 17. $\{\frac{1}{2}, -\frac{1}{2}, 2, -2\}$.
19. $\{256\}$. 21. $\{1\}$. 23. $\{343, -125\}$. 25. $\{8, 7\}$. 27. $\{3\}$.
29. $\{-3, -2, 2, 3\}$.

EXERCISES 9.7

1. $\{x \mid x > 3 \text{ or } x < -3\}$.

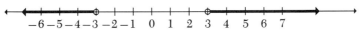

3. $\{x \mid x > 2 \text{ or } x < -2\}$.

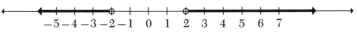

5. $\{x \mid x \geq 5 \text{ or } x \leq -5\}$.

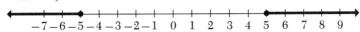

7. $\{x \mid -3 \leq x \leq 3\}$.

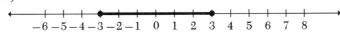

9. $\{x \mid x < -3 \text{ or } x > 2\}$.

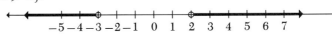

11. $\{x \mid x \leq -2 \text{ or } x \geq 3\}$.

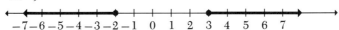

13. $\{x \mid x > 1 \text{ or } x < -3\}$.

15. $\{x \mid x > 2 \text{ or } x < 1\}$.

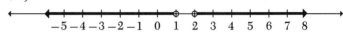

17. $\{x \mid x \geq 3 \text{ or } x \leq 2\}$.

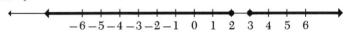

19. $\{x \mid \frac{3}{2} < x < 2\}$.

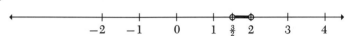

CHAPTER 9   REVIEW

1. $\{4, -4\}$.   **3.** $\{3, 10\}$.   **5.** $\{\frac{7}{4}, -2\}$.   **7.** $\{-4, -\frac{2}{5}\}$.
9. $\{3 + i, 3 - i\}$.   **11.** $\{0, \frac{7}{2}\}$.   **13.** $\{33\}$.   **15.** $\{-1, -2\}$.
17. $\{2, -2, \frac{1}{2}, -\frac{1}{2}\}$.   **19.** $\{x \mid -2 < x < 3\}$.

EXERCISES 10.1

1.
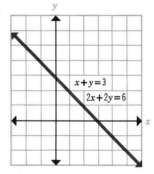
Dependent

3.
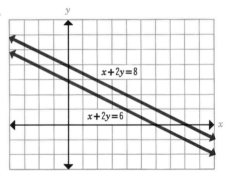
Inconsistant

5.
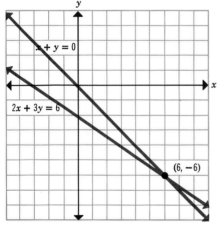
Independent

7.
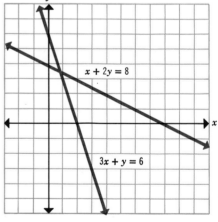
Independent

9.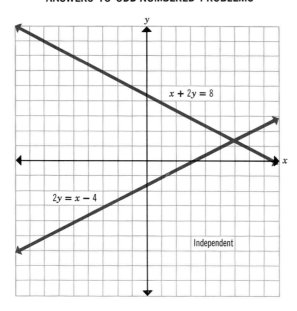

EXERCISES 10.2

1.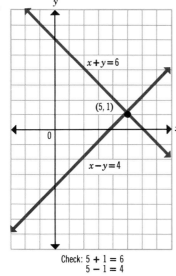

Check: $5 + 1 = 6$
$5 - 1 = 4$

3.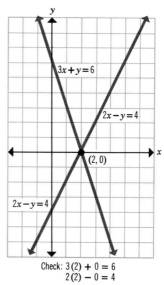

Check: $3(2) + 0 = 6$
$2(2) - 0 = 4$

5.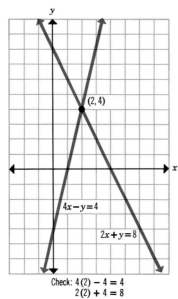
Check: $4(2) - 4 = 4$
$2(2) + 4 = 8$

7.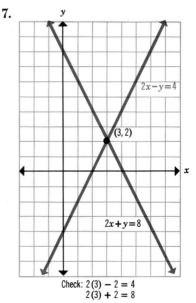
Check: $2(3) - 2 = 4$
$2(3) + 2 = 8$

9.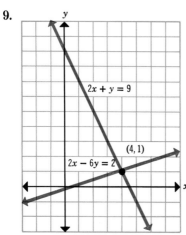
Check: $2(4) - 5(1) = 2$
$2(4) + 1 = 9$

11.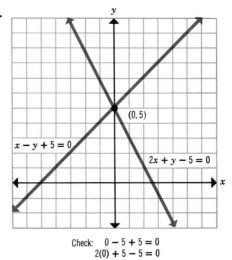
Check: $0 - 5 + 5 = 0$
$2(0) + 5 - 5 = 0$

13.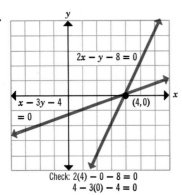
Check: $2(4) - 0 - 8 = 0$
$4 - 3(0) - 4 = 0$

15.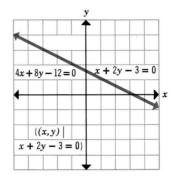

ANSWERS TO ODD-NUMBERED PROBLEMS

EXERCISES 10.3

1. $\{(3, 1)\}$. 3. $\{(2, 3)\}$. 5. $\{(\frac{11}{9}, \frac{4}{3})\}$. 7. $\{(-2, 3)\}$.
9. $\{(x, y) \mid x + 2y = 4\}$. 11. $\{(\frac{107}{47}, -\frac{6}{47})\}$. 13. $\{(2, -1)\}$.
15. $\{(-\frac{7}{11}, \frac{2}{55})\}$. 17. $\{(17, 11)\}$. 19. $\{(3, -3)\}$.

EXERCISES 10.4

1. -4. 3. 1. 5. 2. 7. $\frac{7}{16}$. 9. 0. 11. -12. 13. 0.
15. 14.

EXERCISES 10.5

1. $\{(3, 4)\}$. 3. $\{(\frac{5}{2}, -\frac{1}{2})\}$. 5. $\{(3, 5)\}$. 7. $\{(-19, -29)\}$.
9. $\{(5, 1)\}$. 11. $\{(-2, 1)\}$. 13. $\{(0, 7)\}$. 15. $\{(-\frac{10}{3}, -\frac{2}{3})\}$.
17. $\{(2, -3)\}$. 19. $\{(3, 1)\}$. 21. $\{(\frac{6}{5}, -\frac{17}{5})\}$. 23. $\{(-4, 5)\}$.
25. $\{(\frac{1}{4}, -\frac{1}{4})\}$. 27. $\{(-\frac{29}{9}, \frac{8}{27})\}$. 29. $\{(-1, 3)\}$. 31. $\{(\frac{1}{2}, 4)\}$.
33. $\{(7, -2)\}$. 35. $\{(1, 5)\}$. 37. $\varnothing$. 39. $\{(\frac{37}{38}, \frac{5}{19})\}$.

EXERCISES 10.6

1. $\{(-1, -3, 2)\}$. 3. $\{(\frac{1}{2}, -\frac{7}{6}, -\frac{1}{3})\}$. 5. $\{(\frac{1}{2}, \frac{1}{3}, \frac{1}{4})\}$.

EXERCISES 10.7

1. -18. 3. 21. 5. 6. 7. 86. 9. -13. 11. $-a$.
13. $-x^2y - xy^2$. 15. x^2. 17. $x - y$. 19. 0.

EXERCISES 10.8

1. $\{(\frac{4}{3}, \frac{10}{9}, \frac{22}{9})\}$. 3. $\{(14, 12, 0)\}$. 5. $\{(2, 2, 1)\}$. 7. $\{(-1, \frac{1}{2}, 2)\}$.
9. $\{(7, 4, -5)\}$.

EXERCISES 10.9

1.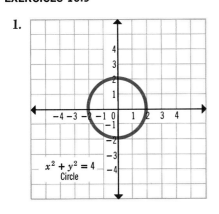

3.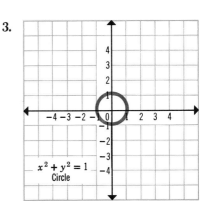

294   ANSWERS TO ODD-NUMBERED PROBLEMS

5.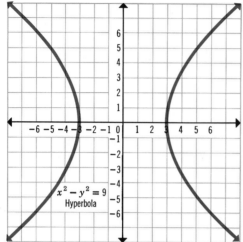
$x^2 - y^2 = 9$
Hyperbola

7.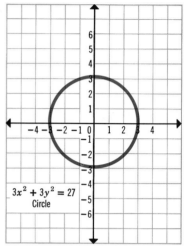
$3x^2 + 3y^2 = 27$
Circle

9.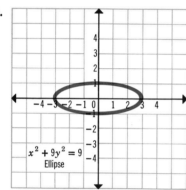
$x^2 + 9y^2 = 9$
Ellipse

11.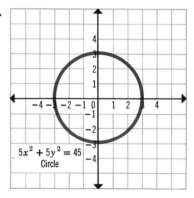
$5x^2 + 5y^2 = 45$
Circle

13.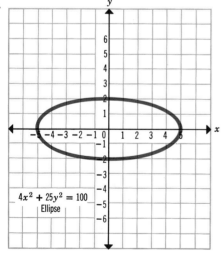
$4x^2 + 25y^2 = 100$
Ellipse

15.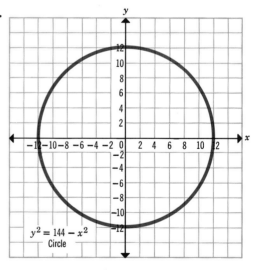
$y^2 = 144 - x^2$
Circle

17.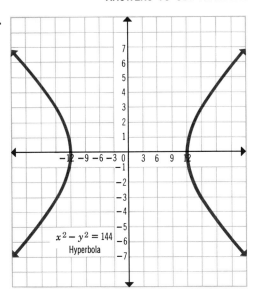

19.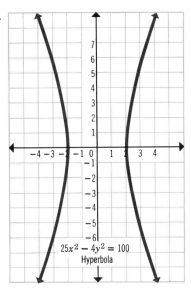

EXERCISES 10.10

1. $\{(4, 0), (-5, 3)\}$.   3. $\{(4, -4), (0, 0)\}$.
5. $\{(-2\sqrt{2}, -2\sqrt{2}), (2\sqrt{2}, 2\sqrt{2})\}$.   7. $\{(0, 2), (2, 0)\}$.
9. $\{(0, 0), (\frac{1}{3}, \frac{1}{3})\}$.   11. $\{(-2, -1), (-2, 1), (2, 1), (2, -1)\}$.
13. $\{(-6, -8), (-6, 8), (6, 8), (6, -8)\}$.
15. $\{(-\frac{2}{3}\sqrt{51}, -\frac{1}{3}\sqrt{6}), (-\frac{2}{3}\sqrt{51}, \frac{1}{3}\sqrt{6}), (\frac{2}{3}\sqrt{51}, \frac{1}{3}\sqrt{6}), (\frac{2}{3}\sqrt{51}, -\frac{1}{3}\sqrt{6})\}$.
17. $\{(-4, -3), (-4, 3), (4, 3), (4, -3)\}$.
19. $\{(-\sqrt{3}, 4), (-\sqrt{3}, -4), (\sqrt{3}, 4), (\sqrt{3}, -4)\}$.

CHAPTER 10   REVIEW

1. $\{(3, 2)\}$.

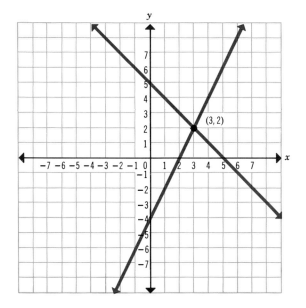

3. $\{(4, -3)\}$.   **5.** 2.   **7.** $\{(-1, 4)\}$.   **9.** $\{(\frac{14}{11}, -\frac{9}{11}, -\frac{5}{11})\}$.
11. 9.   **13.** $\{(-1, 2, 1)\}$.

15.

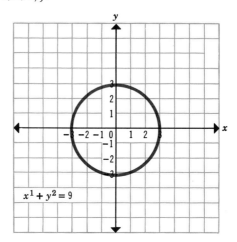

17.

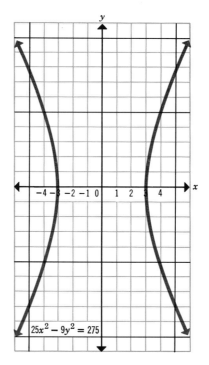

19. $\{(0, 3), (2, 0)\}$.

EXERCISES 11.1

1. 1, 3, 6, 10.   **3.** 1, 4, 9, 16.   **5.** 4, 5, 6, 7.   **7.** $-3, -2, -1, 0$.
9. $\frac{1}{5}, \frac{2}{9}, \frac{3}{13}, \frac{4}{17}$.   **11.** $\frac{1}{2}, \frac{2}{3}, \frac{3}{4}, \frac{4}{5}$.   **13.** 0, 3, 8, 15.   **15.** 1, 7, 17, 31.

17. 0, 1, 64, 6, 561.   **19.** 0, 4, 18, 48.   **21.** $\frac{1}{n}$.   **23.** $\frac{2-n}{2}$ or $1 - \frac{n}{2}$.
25. $(-1)^{n-1}$.

EXERCISES 11.2

1. $4 + 7 + 10 + 13$.   **3.** $2 + 8 + 26 + 80$.
5. $\frac{1}{2} + \frac{1}{3} + \frac{1}{4} + \frac{1}{5} + \frac{1}{6} + \frac{1}{7} + \frac{1}{8}$.   **7.** $7 + 10 + 13 + \cdots + (3n + 4)$.
9. $\frac{1}{2} - \frac{1}{4} + \frac{1}{6} - \frac{1}{8} + \cdots + \frac{(-1)^{n-1}}{2n}$.   **11.** $\sum_{i=1}^{5} i$.   **13.** $\sum_{i=1}^{5} (2i - 1)$.
15. $\sum_{i=1}^{n} \frac{1}{3^i}$.   **17.** $\sum_{i=1}^{n} \frac{1}{i+1}$.   **19.** $\sum_{i=1}^{n} (-1)^{i-1} 2^i$.

EXERCISES 11.3

1. $2n$; 2.   **3.** $-13 + 5n$; 5.   **5.** $18 - 5n$; -5.   **7.** $a + n - 1$; 1.
9. $\frac{3}{4} - \frac{1}{4}n$; $-\frac{1}{4}$.   **11.** 77.   **13.** 332.   **15.** -19.   **17.** 24.
19. 650.   **21.** 2550.   **23.** $d = 2$, $a = 14$.   **25.** 3.   **27.** 129.
29. 9 days.

EXERCISES 11.4

1. Arithmetic.   **3.** Geometric.   **5.** Geometric.   **7.** Geometric.
9. Arithmetic.   **11.** $r = \frac{1}{2}$; $\frac{1}{16}, \frac{1}{32}, \frac{1}{64}$.   **13.** $r = -2$; $-24, 48, -96$.
15. $r = \frac{1}{a^2}$; $\frac{1}{a^5}, \frac{1}{a^7}, \frac{1}{a^9}$.   **17.** $r = a$; $\frac{a^4}{b}, \frac{a^5}{b}, \frac{a^6}{b}$.
19. $r = -\frac{b}{a}$; $-\frac{b^2}{a^2}, \frac{b^3}{a^3}, -\frac{b^4}{a^4}$.   **21.** 768.   **23.** $\frac{1}{243}$.   **25.** $4\sqrt{3}$.
27. -10^{-11}.   **29.** $\frac{1}{27}$.   **31.** 363.   **33.** 1023.   **35.** $\frac{121}{81}$.   **37.** $\frac{5}{8}$.

EXERCISES 11.5

1. $\frac{1}{2}$.   **3.** 3.   **5.** $\frac{4}{5}$.   **7.** $\frac{400}{3}$.   **9.** $-\frac{125}{3}$.   **11.** $\frac{1}{3}$.   **13.** $\frac{1}{6}$.
15. $\frac{145}{33}$.   **17.** $\frac{3136}{999}$.   **19.** $\frac{315}{999}$.   **21.** 42 feet.

EXERCISES 11.6

1. $x^6 + 6x^5y + 15x^4y^2 + 20x^3y^3 + 15x^2y^4 + 6xy^5 + y^6$.
3. $x^5 - 15x^4 + 90x^3 - 270x^2 + 405x - 243$.
5. $a^7 - 14a^6b + 84a^5b^2 - 280a^4b^3 + 560a^3b^4 - 672a^2b^5 + 448ab^6 - 128b^7$.
7. $y^4 + 4\sqrt{2}y^3 + 12y^2 + 8\sqrt{2}y + 4$.
9. $x^2\sqrt{x} - 5x^2\sqrt{y} + 10xy\sqrt{x} - 10xy\sqrt{y} + 5y^2\sqrt{x} - y^2\sqrt{y}$.

11. $a^{10} + 30a^9b + 405a^8b^2 + 3240a^7b^3$.
13. $y^{15} - \frac{15}{4}y^{14} + \frac{105}{16}y^{13} - \frac{455}{64}y^{12}$.
15. $1024x^{10} - 15{,}360x^9y + 103{,}680x^8y^2 - 414{,}720x^7y^3$. **17.** 1680.
19. $\frac{1}{7}$. **21.** $\frac{33}{14}$. **23.** $h^2 + 3h + 2$ or $(h+1)(h+2)$. **25.** $7xy^6$.
27. $3360x^{12}$. **29.** 59136.

CHAPTER 11 REVIEW

1. (a) $-5, -9, -13$; (b) $-\frac{11}{2}, -8, -\frac{21}{2}$;
   (c) $1 - 5\sqrt{2}, 1 - 7\sqrt{2}, 1 - 9\sqrt{2}$. **3.** (a) 39; (b) 165.
5. (a) $\frac{4}{729}$; (b) $\frac{13120}{729}$. **7.** \$71,750. **9.** (a) $\frac{4}{3}$; (b) $\frac{1}{9}$.
11. $a^5 + 5a^4b + 10a^3b^2 + 10a^2b^3 + 5ab^4 + b^5$.
13. $\dfrac{1}{729}x^6 - \dfrac{2}{81}x^4 + \dfrac{5}{27}x^2 - \dfrac{20}{27} + \dfrac{5}{3}\dfrac{1}{x^2} - 2\dfrac{1}{x^4} + \dfrac{1}{x^6}$.
15. $x^{12} + 12x^{11}y + 66x^{10}y^2 + 220x^9y^3$.
17. $x^{20} - 20x^{18} + 190x^{16} - 1140x^{14}$. **19.** $-6435x^9$.

EXERCISES 12.1

1.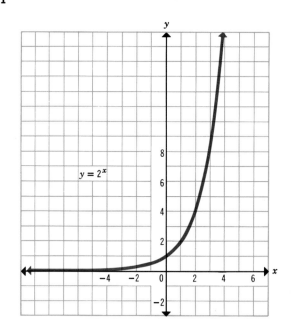

$y = 2^x$

3.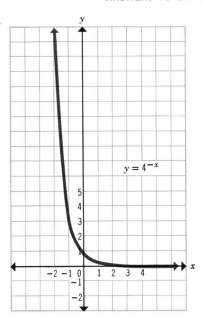

5.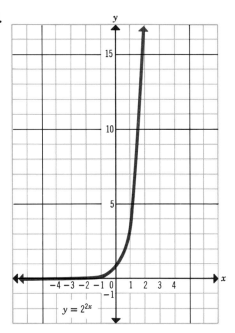

7.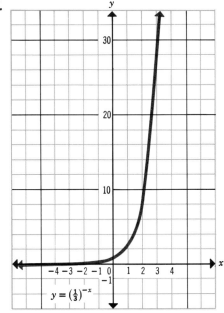

9.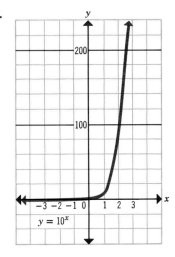

11. (a) Increases; (b) increases; (c) decreases; (d) decreases; (e) decreases; (f) increases. **13.** $2^{3x} = (2^3)^x = 8^x$.

ANSWERS TO ODD-NUMBERED PROBLEMS

EXERCISES 12.2

1. $\log_2 4 = 2$. **3.** $\log_{10} 100 = 2$. **5.** $\log_2 \frac{1}{2} = -1$. **7.** $\log_{100} 10 = \frac{1}{2}$. **9.** $\log_{10} 0.01 = -2$. **11.** $\log_{10} 1000 = 3$. **13.** $2^6 = 64$. **15.** $4^0 = 1$. **17.** $(\frac{1}{2})^{-3} = 8$. **19.** $8^{2/3} = 4$. **21.** 2. **23.** 3. **25.** -2. **27.** 1.
29.

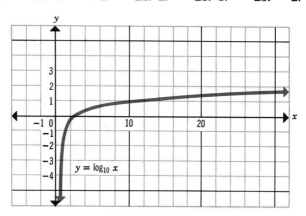

EXERCISES 12.3

1. $\log_a xy$. **3.** $\log_a x^3$. **5.** $\log_a xy^2$. **7.** $\log_a (xy)^{1/4}$. **9.** $\log_a \dfrac{x^3}{y^{4/3}}$.
11. 0.778. **13.** 0.954. **15.** -0.359. **17.** 0.223. **19.** 0.176.

EXERCISES 12.4

1. 0.6274. **3.** 0.8561. **5.** 0.0000. **7.** 0.3324. **9.** 0.5198.
11. 1.9415. **13.** 2.0453. **15.** 2.3655. **17.** $9.7033 - 10$.
19. 0.2279. **21.** $7.8494 - 10$. **23.** 5.3747. **25.** $7.6304 - 10$.
27. 5.9939. **29.** 7.9689. **31.** $9.5038 - 10$. **33.** $6.6946 - 10$.
35. 1.8681. **37.** $8.9031 - 10$. **39.** $7.4843 - 10$. **41.** 7.0374.
43. $8.6191 - 10$. **45.** $2.4409 - 10$. **47.** 9.45. **49.** 9.41.
51. 1.01. **53.** 79.4. **55.** 54,400. **57.** 643. **59.** 0.307.
61. 0.0402. **63.** 0.0000208. **65.** 0.249. **67.** 0.539.
69. 0.00000475. **71.** 146,000. **73.** 0.0462. **75.** 4,140,000.

EXERCISES 12.5

1. 0.1559. **3.** 0.4924. **5.** 2.2716. **7.** 3.5859. **9.** $8.9010 - 10$.
11. $7.9014 - 10$. **13.** 1.8975. **15.** 3.1421. **17.** $7.0457 - 10$.
19. $7.4290 - 10$. **21.** 2.222. **23.** 29.44. **25.** 128.3. **27.** 85,280.
29. 362,200. **31.** 0.1374. **33.** 0.01398. **35.** 0.001646.
37. 0.001007. **39.** 0.001298.

ANSWERS TO ODD-NUMBERED PROBLEMS

EXERCISES 12.6

1. 51.49. **3.** 5.329. **5.** 5.176. **7.** 2.545. **9.** 19.74.
11. 0.02997. **13.** 8.650. **15.** 1.582. **17.** 0.001624. **19.** 1.318.
21. 21.74. **23.** 1134. **25.** 2.358. **27.** 14.44. **29.** 4.434.
31. 429.4. **33.** 2.563. **35.** 0.000000001995. **37.** 0.4104.
39. -0.006045. **41.** \$12,070. **43.** 606 cubic inches.

EXERCISES 12.7

1. $\{1.161\}$. **3.** $\{4.585\}$. **5.** $\{0.6108\}$. **7.** $\{2.483\}$. **9.** $\{-4.861\}$.
11. $\{25{,}000\}$. **13.** $\{63\}$. **15.** $\{\frac{1}{6}\}$ or $\{0.166\ldots\}$. **17.** $\{47.5\}$.
19. $\{-1.444\}$.

CHAPTER 12 REVIEW

1.

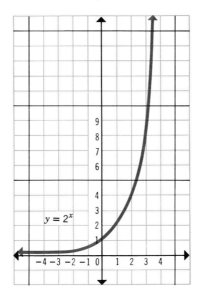

3. (a) $3^4 = 81$; (b) $10^{-3} = 0.001$.
5. (a) 37.6; (b) 0.931; (c) 6,410; (d) 0.0128.
7. $3\log_{10} x - \log_{10} 3 - \log_{10} y$. **9.** $\frac{1}{5}\log_{10} x$.
11. $\log_{10} 3 + 2\log_{10} x + \log_{10} y - \frac{3}{4}\log_{10} z$. **13.** 6.526. **15.** 13.17.
17. $\{-0.553\}$. **19.** $\{1\}$.

INDEX

Abscissa, 116
Absolute value, 9
Addition, associative property of, 7
   closure property of, 7
   commutative property of, 7
   of complex numbers, 139–140
   identity element of, 7, 9, 140
   of polynomials, 49–51
   property of equality, 5
   property of inequality, 10
   of rational expressions, 71–75
   of real numbers, 14–17
Additive cancellation property of equality, 5
Additive identity, 7, 9, 140
Additive inverse, 7, 50, 140
Algebraic expression, 46
Answers to odd-numbered problems, 265–301
Antilogarithm, 250–257
Arithmetic progression, 223–225
   common difference of, 223
   nth term of, 224
Arithmetic series, 225–226
   sum of, 225
Associative property, 139–140
   of addition, 7
   generalized, 9
   of multiplication, 7, 143

Axioms, 7–8
Axis of abscissas, 111
Axis of ordinates, 111

Base, 25
Binomial, 47
Binomial expansion, 235–237
   rth term of, 235
Braces, 1

Cartesian coordinate system, 115–117
Cartesian plane, 111
Cartesian product, 109–115
Characteristic of logarithm, 250
Circle, 209
Closure property, of addition, 7
   of multiplication, 7
Coefficient, 46
Common difference, 223
Common logarithms, 248–254
   table of, 261–263
Commutative property, of addition, 7, 139
   of multiplication, 7, 142
Completing the square, 157–162
Complex numbers, 138–153
   addition of, 139–140
   additive identity of, 140
   additive inverse of, 140

Complex numbers, conjugate of, 149–150
  division of, 144–146
  imaginary component of, 138
  multiplication of, 142–144
  multiplicative identity of, 143
  multiplicative inverse of, 144–145
  purely real, 148
  real components of, 138
  square roots which are, 151–152
  standard form, 147–149
  subtraction of, 140–141
Complex rational expressions, 82–84
Composite numbers, 13–14
Conditional equations, 95
Conics, 209–213
Conjugate, 149–150
Constant function, 242
Constant of variation, 132
Coordinate system, 112, 115–117
Counting numbers, 2, 3

Determinants, 196–197
  elements of, 196
  expansion of, 204
  minors, 203
  principal diagonal, 196
  secondary diagonal, 196
  second order, 196
  sign array of, 204
  third order, 203–205
Difference, 14
Direct variation, 132–135
Discriminant, 174
Distributive property, 8, 143
  generalized, 10, 60
Dividend, 20
Division, of complex numbers, 144–146
  of polynomials, 57–58
  of rational expressions, 79–80
  of real numbers, 19–21
  by zero, 20–21
Divisor, 20
Domain, 2, 88, 122

Elementary transformation, 89
Element of a set, 1
Ellipse, 211
Empty set, 2
Equality relation, 4
  properties of, 4–5, 88
Equal sets, 2
Equation, 5, 87–108
  algebraic, 87–88
  applications of, 101–107
  conditional, 95
  containing radicals, 181–182
  equivalent, 88–89
  exponential, 257–259
  first degree algebraic, 88, 90
  fractional, 179–180
  identical, 95
  linear, 88, 188
  logarithmic, 257–259
  quadratic, 168–187
  with quadratic form, 182–184
  solution of, 88
  solution set of, 88
  solving, 89–92
  solving for specified symbols, 93–94
Equivalence relation, 4
Equivalent equation, 88–89
Exponential equations, 257–259
Exponential function, 241–244
Exponents, 25–45
  laws of positive integer, 25–29, 32–33
  negative rational number, 35–36
  positive rational number, 32–33
  zero, 35

Factoring, 59–67
  binomials, 62–65

grouping by pairs, 60–61
sum and difference of two cubes, 66–67
trinomials, 62–65
Factors, 13
Field, 8
Finite sequence, 219
Finite series, 221
Finite set, 3
First degree algebraic equations, 88
Fraction, 68
three signs of, 75
Fractional equations, 179–180
Fundamental theorem of arithmetic, 13
Function, 109–137
constant, 242
exponential, 241–244
linear, 128–132, 154
logarithmic, 244–246
notation, 125–127
quadratic, 154–167
sequence, 219
value of, 127
vertical line test, 125
Functional notation, 126–127

Geometric progressions, 227–228
common ratio of, 227
nth term of, 228
Geometric series, 229–231
infinite, 232–234
Graphs of linear functions, 128–132
Graphs of quadratic functions, 155–157

Hyperbola, 212

Identities, 95–96
Identity element, of addition, 7, 9, 140
of multiplication, 7
Index, 31, 222
Inequalities, 97–101
application of, 101–107

fundamental properties of, 97
graph of solution set, 97
quadratic, 168–187
solution of, 97
solution set of, 97
Infinite geometric series, 232–234
sum of, 232
Infinite sequence, 219
Infinite set, 3
Integers, 3, 13
Inverse, of addition, 7
of multiplication, 7, 10
Inverse variation, 132–135
Irrational numbers, 3, 4, 6

Laws of exponents, 25–29
Laws of logarithms, 246–247
Least common denominator, 72
Least common multiple, 70–71
Limit, 232
Linear equation, 88, 188
Linear function, 128–132
Linear interpolation, 252–254
Logarithm, base of, 244
base, 10, 248–254
characteristic of, 250
common, 248–254
laws of, 246–247, 255
mantissa of, 250
table of, 261–263
Logarithmic computation, 255–256
Logarithmic equations, 257–259
Logarithmic function, 244–245
graph of, 244–245
Log table, 248–250, 261–263

Mantissa, 250
Member of equation, 5
Member of set, 1
Monomial, 47
Multiplication, associative property of, 7

Multiplication, closure property of, 7
  commutative property of, 7
  of complex numbers, 140–141
  identity element of, 7, 9, 143
  of polynomials, 52–53
  property of equality, 5
  property of inequality for negative numbers, 11
  property of inequality for positive numbers, 10
  property of zero, 7
  of rational expressions, 77–78
  of real numbers, 19–21
Multiplicative cancellation property of equality, 5
Multiplicative identity, 7, 9, 143
Multiplicative inverse, 7, 10, 144–145

$n!$, 236
Natural numbers, 2, 3
Number(s), complex, 138–153
  composite, 13–14
  counting, 2, 3
  field, 8
  integers, 3, 13
  irrational, 3, 4, 6
  natural, 2, 3
  prime, 13–14, 70
  rational, 3, 6
  real, 4, 6–12
  whole, 3
Number field, 8

Open sentences, 87
  in two variables, 117–121
Opposite, 7
  of polynomial, 50
Ordered pair, 109
  components of, 109
  equal, 109
Ordered triple, 201
Order relation, 10–11

Ordinate, 116
Origin, 116

Parabola, 155, 159, 209
  vertex of, 160–161
Perfect squares, 62
Polynomials, 46–47
  addition of, 49–51
  additive inverse of, 50
  classification of, 47–48
  degree of, 47
  division of, 57–58
  evaluating, 48
  factoring, 59–67
  opposite of, 50
  over the integers, 47
  multiplication of, 52–53
  subtraction of, 49–51
  terms of, 46
  value of, 48
  vanishes, 68
  zero, 68–69
Power, 25
Prime numbers, 13–14, 70
Principal square root, 30
Product, 19
Proportion, 134–135

Quadrant, 116
Quadratic equations, 168–187
  properties of roots of, 176–178
  solution by factoring, 170–172
  standard form of, 168
Quadratic form, 183
Quadratic formula, 173–175
Quadratic functions, 154–167
  applications of, 162–165
  graphs of, 155–157
  maximum value of, 155
  minimum value of, 155
Quadratic inequalities, 168–187
Quotient, 20

Radical(s), 30–32
　adding, 41–42
　changing form of, 37–40
　equations containing, 181–182
　index of, 31
　multiplication of, 41–42
　operations containing, 41–42
　order of, 41
　quotient of, 42
　rationalizing the denominator, 38
　similar, 41
　simplest form of, 38
　simplifying, 37–40
　subtraction of, 41–42
Radicand, 31
Rational expression, 68–86
　addition of, 71–75
　additive inverse of, 74–75
　complex, 82–84
　division of, 79–80
　multiplication of, 77–78
　simplification of, 68–69
　subtraction of, 71–75
Rationalizing the denominator, 38
Rational number exponents, 32–33, 35–36
Rational numbers, 3, 6
Real components of complex numbers, 138
Real number field, 8
Real numbers, 4
　absolute value of, 8
　addition of, 14–17
　axioms, 7–8
　difference of, 14
　division of, 19–21
　multiplication of, 19–21
　and their properties, 6–12
　subtraction of, 14–17
Real plane, 115
Reciprocal, 7, 10
Rectangular coordinate system, 116

Reflexive property, 4
Relation(s), 109–137
　equality, 4
　equivalence, 4
　domain of, 122
　range of, 122
　roster method of denoting, 122
Replacement set, 2, 87–88
Roots, 30–32

Scientific notation, 44, 249
Sentences, open in two variables, 117–121
　solution of, 118
　solution set of, 118
Sequence function, 219
Sequences and series, 219–240
　arithmetic progression, 223–225
　finite, 219
　function, 219
　general term of, 219
　geometric progression, 227–228
　infinite, 219
　terms of, 219
Series, 221–222
　arithmetic, 225–226
　finite, 221
　geometric, 229–231, 232–234
Set, 1
　belonging to, 1
　builder notation, 2
　element of, 1
　elements listed, 1
　empty, 2
　equal, 2
　finite, 3
　infinite, 3
　member of, 1
　replacement, 2
　subset of, 2
Slope, 129
Solution, 88, 118

Solution set, 88, 118, 188
Special products, 54–56
Square root, 30, 151–152
  principal, 30
Standard form of complex number, 147–149
Statement, 87
Subset, 2
Substitution property of equality, 4, 5
Subtraction, of complex numbers, 140–141
  of polynomials, 49–51
  property of equality, 5
  of rational expressions, 71–75
  of real numbers, 14–17
Summand, 222
Summation notation, 222
Summation sign, 222
Symbols, 1
Symmetric property, 4
Systems of equations, 188–218
  analytic solution of, 193–195, 201–202
  dependent, 188–190
  inconsistent, 188–190
  independent, 188–190
  one linear and one quadratic, 213–214
  solution by determinants, 197–200, 206–209
  solution by graphing, 190–192
  two second-degree, 214–216

Term of a sequence, 219
Transitive property, 4, 10, 97
Tree diagram, 110
Trichotomy property, 10
Trinomial, 47

Variable, 2
Variation, constant of, 132
  direct, 132–135
  inverse, 132–135
  joint, 134
Vertical line test, 125

Whole numbers, 3

Zero, and division, 20–21
  exponent, 35
  multiplication property of, 7